Advances in

ECOLOGICAL RESEARCH

VOLUME 36

Advances in Ecological Research

Series Editor: HAL CASWELL

Biology Department
Woods Hole Oceanographic Institution
Woods Hole, Massachussetts

Advances in
ECOLOGICAL
RESEARCH

VOLUME 36

Food Webs: From Connectivity to Energetics

Edited by
HAL CASWELL

Biology Department
Woods Hole Oceanographic Institution
Woods Hole, Massachussetts

2005

ELSEVIER
ACADEMIC
PRESS

AMSTERDAM • BOSTON • HEIDELBERG • LONDON
NEW YORK • OXFORD • PARIS • SAN DIEGO
SAN FRANCISCO • SINGAPORE • SYDNEY • TOKYO

Elsevier Academic Press
525 B Street, Suite 1900, San Diego, California 92101-4495, USA
84 Theobald's Road, London WC1X 8RR, UK

This book is printed on acid-free paper.

For all information on all Elsevier Academic Press publications
visit our Web site at www.books.elsevier.com

ISBN: 0-12-013936-7

PRINTED IN THE UNITED STATES OF AMERICA
05 06 07 08 9 8 7 6 5 4 3 2 1

Contributors to Volume 36

STEPHEN R. CARPENTER, *Center for Limnology, 680 North Park Street, University of Wisconsin, Madison, Wisconsin 53706, USA. e-mail: srcarpen@facstaff.wisc.edu*

JOEL E. COHEN, *Laboratory of Populations, Rockefeller and Columbia Universities, Box 20, 1230 York Avenue, New York, New York 10021, USA. e-mail: cohen@rockefeller.edu*

ALAN G. HILDREW, *School of Biological Sciences, Queen Mary and Westfield College, University of London, Mile End Road, London E1 4NS, United Kingdom. e-mail: a.hildrew@qmw.ac.uk*

TOMAS JONSSON, *Department of Natural Science, University of Skövde, Skövde, Sweden. e-mail: tomas.jonsson@inv.his.se*

DANIEL C. REUMAN, *Rockefeller University, Box 20, 1230 York Avenue, New York, New York 10021, USA. e-mail: reumand@rockfeller.edu*

DOUGIE C. SPEIRS, *Department of Statistics and Modelling Science, University of Strathclyde, Glasgow G1 1XH, United Kingdom. e-mail: dougie@stams.strath.ac.uk*

GUY WOODWARD, *Department of Zoology, Ecology and Plant Science, University College Cork, Ireland. e-mail: g.woodward@ucc.je*

Preface

In 1923, V. Summerhayes and Charles Elton published an account of the ecology of the arctic island of Spitsbergen. Their report included observations of the food habits of many of the species, summarized in a diagram showing feeding relationships. Four years later, Elton included this diagram in his influential ecology textbook (Elton, 1927). He called it a *food-cycle*; today it would be known as a food web (indeed, it appears as the earliest published web in the compilation of Cohen *et al.* (1990)).

The Summerhayes-Elton web has some properties common to many webs published since. Some of its components are species (the purple sandpiper, the polar bear), others are higher taxonomic groupings (algae, hymenopterans, protozoa), and still others are non-taxonomic ecological groupings ("marine animals"). The web includes no information on any properties of the species (their abundance, for example). It documents feeding links, but says nothing about which links are more important and which are less.

Such *connectance webs* (see Woodward *et al.*, this volume, for terminology) document genuine ecological properties of a community (the long-tailed duck feeds on freshwater invertebrates; the puffin does not). But they are silent on many important ecological patterns that can only be addressed with a *quantified web*, one that includes some information on the strength of the links in the web. [Elton was perfectly well aware of this, and included discussions of the importance of body size in food web structure and hinted at relationships between body size patterns and patterns of population oscillations (Elton, 1927).]

This volume of *Advances in Ecological Research* presents three studies that, individually and collectively, make significant contributions to the analysis of food webs, because they integrate information on body size, abundance, and productivity with the pattern of connectivity. To put these chapters in context, it is helpful to recall several different ways in which ecologists have traditionally thought of food webs.

Food webs as wiring diagrams. Connectance webs are like wiring diagrams; they show only the most basic topological structure of the food web. Even so, they provide information on the density of links, the distribution of the numbers of predators and prey with which a species is associated, the classification of species into trophic levels, the length of food chains within the web, and many other properties. In a long series of papers, Cohen and his collaborators have explored these patterns, particularly in the context of the *cascade model* (Cohen *et al.*, 1990; see also the chapter by Jonsson *et al.* in this volume).

The cascade model is not a model of how food webs function, or of how they came to have the structure they do. Instead, it is a stochastic recipe for constructing connectance webs, given a set of species and the value for the model's single free parameter (the density of feeding links). It is as if food webs were constructed by elves in a workshop at the North Pole. If each elf was equipped with a copy of the cascade model and a random number generator, then the connectance webs of the world would display certain statistical properties. Given sufficient mathematical skill, these properties can be calculated from the model (Cohen *et al.*, 1990), and then compared with the properties of documented connectance webs (including the Summerhayes-Elton web).

The agreement between the model and the published webs is, by and large, good, though not perfect. Just how good is not the issue here. What is important is that food webs are not constructed by elves (I don't think so, anyway). Thus, any agreement between real food web wiring diagrams and the predictions of the cascade model leads directly to the question: What real ecological factors or processes might lead to food webs with these properties? One candidate often suggested is body size. Predators are often larger than their prey, and this fact might help produce the kind of organization implied by the cascade model.

The chapters in this volume quantify connectance webs in terms of body size and abundance of the component species. This produces a plethora of trivariate (food web, body size, and abundance), bivariate, and univariate relationships, and leads to new ways of diagramming and characterizing food webs. The contributions here are based on some of the most detailed and intensive food web sets ever collected, and document new food web patterns, the implications of which are still to be explored.

Food webs as plumbing. Food webs are not only wiring diagrams showing which species are connected to which. They are pathways – plumbing of sorts – through which matter and energy flow within ecosystems. Accounting for these fluxes is a major task in studies of nutrient cycling and productivity.

Such studies require not only the connectance web, but information on the *rates* of production and consumption characterizing each of the links. This is a far more onerous requirement than simply quantifying a web in terms of body size. It may explain why, although studies of connectance webs are often criticized because of the aggregation of species within the webs, the webs used in studies of nutrient cycling are often much more highly aggregated (e.g., the widely used plankton model of Fasham *et al.* (1990), which contains only seven highly aggregated components).

The chapter by Woodward *et al.* combines measurements of ingestion and productivity (the latter extrapolated from other, similar ecosystems) to estimate fluxes and uses those estimates to characterize interaction strength. Reuman and Cohen also estimate fluxes, but they do so by using allometric scaling relationships with body size. Their approaches hold forth the intriguing link of a connection between food web theory and recent advances in metabolic ecology (Brown *et al.*, 2004).

Food webs as clockworks. Each species in a food web changes in abundance at a rate that depends, inter alia, on the rate at which it consumes resources and the rate at which it is consumed by its predators. Thus, the entire food web is a dynamic system and can be analyzed as such. Ecologists, at least as far back as Elton, have worried about the dynamic consequences of food web organization, and studies of such dynamic properties as stability, resilience, reactivity, and permanence are still common.

A dynamic model requires even more information: not only the connectance pattern and the fluxes, but the functions that tell how the fluxes are determined by the species abundances and the environment. The dynamic consequences of the patterns discovered in this volume remain to be explored.

Understanding the structure and function of food webs is a central problem in ecology. We hope that this volume will contribute to new analyses, new theories, and new data collection efforts.

REFERENCES

Brown, J.H., Gillooly, J.F., Allen, A.P., Savage, V.M. and West, G.B. (2004) Toward a metabolic theory of ecology. *Ecology* **85**, 1171–1789.
Cohen, J.E., Briand, F. and Newman, C.M. (1990) *Community Food Webs: Data and Theory.* Springer-Verlag, Berlin, Germany.
Elton, C.S. (1927) *Animal Ecology.* MacMillan, New York, New York.
Fasham, M.J.R., Ducklow, H.W. and McKelvie, S.M. (1990) A nitrogen-based model of plankton dynamics in the oceanic mixed layer. *J. Mar. Res.* **48**, 591–639.
Summerhayes, V.S. and Elton, C.S. (1923) Contributions to the ecology of Bear Island and Spitsbergen. *J. Ecol.* **11**, 214–286.

Hal Caswell
Biology Department MS-34
Woods Hole Oceanographic Institution
Woods Hole, MA 02543

Contents

Food Webs, Body Size, and Species Abundance in Ecological Community Description

TOMAS JONSSON, JOEL E. COHEN, AND
STEPHEN R. CARPENTER

Quantification and Resolution of a Complex, Size-Structured Food Web

GUY WOODWARD, DOUGIE C. SPEIRS, AND ALAN G. HILDREW

**Estimating Relative Energy Fluxes Using the Food Web, Species Abundance,
and Body Size**

DANIEL C. REUMAN AND JOEL E. COHEN

Food Webs, Body Size, and Species Abundance in Ecological Community Description

TOMAS JONSSON, JOEL E. COHEN, AND
STEPHEN R. CARPENTER

ADVANCES IN ECOLOGICAL RESEARCH VOL. 36
0065-2504/05 $35.00

I. SUMMARY

This chapter demonstrates that methods to describe ecological communities can be better understood, and can reveal new patterns, by labeling each species that appears in a community's food web with the numerical abundance and average body size of individuals of that species. We illustrate our new approach, and relate it to previous approaches, by analyzing data from the pelagic community of a small lake, Tuesday Lake, in Michigan. Although many of the relationships we describe have been well studied individually, we are not aware of any single community for which all of these relationships have been analyzed simultaneously. An overview of some of the results of the present study, with further theoretical extensions, has been published elsewhere (Cohen *et al.*, 2003).

Our new approach yields four major results. Though many patterns in the structure of an ecological community have been traditionally treated as independent, they are in fact connected. In at least one real ecosystem, many of these patterns are relatively robust after a major perturbation. Some of these patterns may be predictably consistent from one community to another. Locally, however, some community characteristics need not necessarily coincide with previously reported patterns for guilds or larger geographical scales.

We describe our major findings under these headings: trivariate relationships (that is, relationships combining the food web, body size, and species abundance); bivariate relationships; univariate relationships; and the effects of food web perturbation.

A. Trivariate Relationships

Species with small body mass occur low in the food web of Tuesday Lake and are numerically abundant. Larger-bodied species occur higher in the food web and are less numerically abundant. Body size explains more of the variation in numerical abundance than does trophic height. Body mass varies almost 12 orders of magnitude and numerical abundance varies by almost 10 orders of magnitude, but biomass abundance (the product of body mass times numerical abundance) varies by far less, about

5 orders of magnitude. The nearly inverse relationship between body mass and numerical abundance, and the relative constancy of biomass, are illustrated by a new food web graph (Fig. 3), which shows the food web in the plane with axes corresponding to body mass and numerical abundance.

B. Bivariate Relationships

The pelagic community of Tuesday Lake shows a pyramid of numbers but not a pyramid of biomass. The biomass of species increases very slowly with increasing body size, by only 2 orders of magnitude as body mass increases by 12 orders of magnitude. The biomass-body size spectrum is roughly flat, as in other studies at larger spatial scales. Prey body mass is positively correlated to predator body mass. Prey abundance and predator abundance are positively correlated for numerical abundance but not for biomass abundance. Body size and trophic height are positively correlated. Body size and numerical abundance are negatively correlated.

The slope of the linear regression of log numerical abundance as a function of log body size in Tuesday Lake is not significantly different from $-3/4$ across all species but is significantly greater than -1 at the 5% significance level. This $-3/4$ slope is similar to that found in studies at larger, regional scales, but different from that sometimes observed at local scales. The slope within the phytoplankton and zooplankton (each group considered separately) is much less steep than $-3/4$, which is in agreement with an earlier observation that the slope tends to be more negative as the range of body masses of the organisms included in a study increases. A novel combination of the food web with data on body size and numerical abundance, together with an argument based on energetic mechanisms, refines and tightens the relationship between numerical abundance and body size.

The regression of log body mass as a linear function of log numerical abundance across all species has a slope not significantly different from -1, but significantly less than $-3/4$. The estimated slope is significantly different from the reciprocal of the estimated slope of log numerical abundance as a function of log body mass. Thus, if log body mass is viewed as an independent variable and log numerical abundance is viewed as a dependent variable, the slope of the linear relationship could be $-3/4$ but could not be -1 at the 5% significance level. Conversely, if log numerical abundance is viewed as an independent variable and log body mass as a dependent variable, the slope of the linear relationship could be -1 but could not be $-4/3$ (which is the reciprocal of $-3/4$) at the 5% significance level. While a linear relationship is a good approximation in both cases, Cohen and Carpenter (in press) showed that only the model with log body mass as the independent variable meets the assumptions of linear regression analysis for these data.

C. Univariate Relationships

The food web of Tuesday Lake has a pyramidal trophic structure. The number of trophic links between species in nearby trophic levels is higher than would be expected if trophic links were distributed randomly among the species. Food chains are shorter than would be expected if links were distributed randomly. Species low in the food web tend to have more predators and fewer prey than species high in the web. The distribution of body size is right-log skewed. The rank-numerical abundance relationship is approximately broken-stick within phytoplankton and zooplankton while the rank-biomass abundance relationship is approximately log-normal across all species. The slope of the right tail of the body mass distribution is much less steep than has been suggested for regional scales and not log-uniform as found at local scales for restricted taxonomic groups.

D. Effect of Food Web Perturbation

The data analyzed here were collected in 1984 and 1986. In 1985, three species of planktivorous fishes were removed and one species of piscivorous fish was introduced. The data reveal some differences between 1984 and 1986 in the community's species composition and food web. Most other community characteristics seem insensitive to this major manipulation.

Different fields of ecology have focused on different subsets of the bivariate relationships illustrated here. Integration of the relationships as suggested in this chapter could bring these fields closer. The new descriptive data structure (food web plus numerical abundance and body size of each species) can promote the integration of food web studies with, for example, population biology and biogeochemistry.

II. INTRODUCTION

An *ecological community* is a set of organisms, within a more or less defined boundary, that processes energy and materials. There are many different notions of an ecological community and many approaches to describing and understanding community structure and function (Paine, 1980; May, 1989). Here we integrate some of these approaches.

A *food web* lists the kinds of organisms in a community and describes which kinds of organisms eat which other organisms. The food web approach (e.g. Cohen, 1989; Lawton, 1989) tries to understand the community through a detailed study of the trophic interactions among the species within the community. Sometimes, it focuses on the population dynamic effects of species on each other (e.g. Pimm, 1982).

The pattern catalog approach tries to understand communities through patterns in the distribution of species characteristics in different communities and under different circumstances. For example, rank-abundance relations, body size distributions, abundance-body size allometry, and biomass spectra are all examples of community characteristics that emerge from species characteristics. How the trophic relations among the species affect these patterns and vice versa has largely been ignored.

In this chapter, we integrate these different approaches. We augment a traditional food web with information on two species characteristics, body size, and abundance, without presenting or testing a particular theory of community organization. Instead, we advocate the idea that many previously studied relationships and distributions can be better understood by connecting the food web with species abundance and body size.

This approach will be illustrated and tested by data on the pelagic community of Tuesday Lake, a small lake in Michigan, in 1984 and 1986. In 1985, the lake was subjected to a major perturbation (see Section IV.A): the three incumbent fish species were removed and a new fish species was introduced. The manipulation significantly affected a number of parameters (e.g., primary production, chlorophyll concentration, zooplankton biomass; Carpenter and Kitchell, 1988). Until the present analysis, the effects of the manipulation on community characteristics, such as the distributions of body size and abundance or the relationship between them, were unknown. We analyze how the perturbation affected several community-level patterns.

Cohen (1991) suggested that body size and abundance of the species in a community could be related to a ranking of the body size of the species by simple allometric or exponential functions. If this relation is confirmed by empirical data, it raises the possibility of predicting a large number of community patterns using only a few input variables. For example, the distributions of body size and abundance in a community could then be approximated from a single variable, the number of species, and a small number of coefficients. Using the data of Tuesday Lake, we demonstrate the existence of simple relationships that could be tested in other communities. If these relationships are subsequently found to hold in general, they could then be used to predict the structure of additional ecological communities.

Many studies of relationships among species characteristics have focused on geographical scales other than that of the local ecosystem. For example, the body size-abundance relationship is often studied using data from a large set of communities (e.g. Damuth, 1981). Such studies are hampered by a lack of information on the ecological constraints operating on species within a particular local community because the studies average data over several communities. Other studies have focused on particular taxa or guilds within a community. This focus reduces the number of species, range of body sizes, or range of trophic levels included when compared to a whole community.

The present study combines data on virtually all the nonmicrobial pelagic species of Tuesday Lake. The organisms, from phytoplankton to fish, span approximately 12 orders of magnitude in body mass and up to 10 orders of magnitude in numerical abundance. We compare some community characteristics in the local community of Tuesday Lake with previously reported patterns for specific taxa or larger geographic scales.

This chapter is not primarily about Tuesday Lake. Others have described Tuesday Lake in much more detail (e.g. Carpenter and Kitchell, 1988, 1993a). Rather, we use Tuesday Lake to illustrate how many previously unrelated descriptions of communities can be brought together (Table 1). The main theme of the chapter is that when data on body size and abundance are associated with each species in a food web, then the community-wide distributions of body size, abundance, and feeding relations become

Table 1 Descriptions of an ecological community that combine information on the food web, body size, and abundance (number of individuals or biomass)

Distributions and relationships analyzed	Food web	Body size	Abundance	Section discussed
Food web statistics The distribution of trophic links The distribution of chain lengths Trophic generality and vulnerability	✔			V.C.1
The distribution of body size Rank-body size		✔		V.C.2
The distribution of numerical and biomass abundance Rank-abundance			✔	V.C.3
Predator-prey body size allometry Body size vs. trophic height Trophic generality and vulnerability vs. body size	✔	✔		V.B.1
Abundance-body size allometry Abundance-body size spectrum Diversity, body size and abundance		✔	✔	V.B.2
Predator-prey abundance allometry Abundance vs. trophic height Ecological pyramids Trophic generality and vulnerability vs. abundance	✔		✔	V.B.3
Trophic position, body size and abundance	✔	✔	✔	V.A

connected, orderly, and intelligible in new ways. Since the relationship among these three attributes affects many other aspects of an ecological community, awareness of these connections contributes to a better overall understanding of community structure and function.

This chapter is organized as follows. Section II.A presents crucial definitions. Section III presents some theoretical predictions for the relationships among the food web and the distributions of body size and abundance. Section IV describes Tuesday Lake, how the data on the food web, body size, and abundance of the species were collected, and the manipulation in 1985. Section V presents and analyzes the data on Tuesday Lake, including the data from 1984 and 1986 but emphasizing the data of 1984. Section VI compares the data of 1984 and 1986 to see the effects on community patterns of the 1985 perturbation. Section VII discusses limitations in the data and the effect of variability. Section VIII summarizes the new insights gained by an integrated trivariate approach.

A. Definitions

Body mass is the average body mass (kg) of an individual of a species. All individuals are included, not only individuals considered adults. *Numerical abundance* means the concentration of individuals (individuals/m^3). *Biomass abundance* is the total amount of biomass per volume (kg/m^3) of a species. Both numerical abundance and biomass abundance depend crucially on the reference volume of water in which average concentration is estimated. Section IV.B describes how these characteristics were measured for different species in Tuesday Lake. Throughout this chapter, the reference volume of water for both estimates of abundance is the epilimnion, which is roughly equivalent to the photic zone, in Tuesday Lake.

A *basal species* is a species recorded as eating no other species. Usually a basal species is autotrophic, but the absence of evidence that a given species consumes any other species may be due to incomplete observation (for example, of endosymbionts). A *top species* is a species recorded as having no other species as predators or consumers. The absence of evidence that a given species is eaten by any other species may be due to incomplete observation (for example, of parasites inside individuals of the species). An *intermediate species* is a species that consumes at least one other species and is consumed by at least one other species in the web. An *isolated species* is a species that has no other species reported as predators or prey.

A *food chain* $(A, B, C, \ldots, X, Y, Z)$ is an ordered sequence of at least two species $A, B, C, \ldots, X, Y, Z$, where A is a basal species and Z is a top species such that each species (except the last, here denoted Z) is eaten by the next species in the list. The *trophic position* of a species in a food chain is $1+$ the number of species preceding it in the ordered list of species in the chain. For example, in the food chain (A, B, C, ..., X, Y, Z), species A has trophic position 1, species B has trophic position 2, species C has trophic position 3, and the trophic position of Z is equal to the number of species in the list. *Trophic height* is the average trophic position of a species in all food chains of which it is a part. Probably because of large size (due to coloniality and/or spines), a few phytoplankton species were not eaten by the herbivores in Tuesday Lake. These isolated species are left out of some analyses. A *food web* is a collection of cross-linked food chains and sometimes includes, in addition, isolated species. *Connectance* is calculated as $2 \times L/(S^2 - S)$, where L is the number of noncannibalistic links and S is the number of connected (that is, nonisolated) species in a food web. The *unlumped web* of Tuesday Lake refers to the food web describing the trophic interactions among the species listed in Appendices 1A and 2A. In the *trophic-species webs*, species with identical sets of prey and predators are aggregated into *trophic species*. *Linkage density* (d) is the number of links per species (i.e., $d = L/S$). The *trophic vulnerability* (V) and the *trophic generality* (G) of a species are the number of predators and the number of prey, respectively, that species has (Schoener, 1989).

For each consumer species j that eats a nonempty set of resource species R_j, we define the available resource biomass B_j and the available resource productivity Π_j as the sum of the available resource biomass or the available resource productivity, respectively, of each of the resource species eaten by consumer j, that is,

$$B_j = \sum_{i \in R_j} \frac{BA_i}{V_i} = \sum_{i \in R_j} \frac{NA_i \times BM_i}{V_i} \qquad (1)$$

and

$$\Pi_j = \sum_{i \in R_j} \frac{P_i}{V_i} = \sum_{i \in R_j} \frac{NA_i \times BM_i^{3/4}}{V_i}. \qquad (2)$$

The available biomass abundance of a resource species i is calculated as the total biomass abundance BA_i of species i divided by the trophic vulnerability V_i, that is, number of consumer species that the resource species i has (including, of course, consumer species j). The available productivity of a resource species i is calculated as the total productivity P_i of species i divided by V_i. The total productivity (kg $\times$ year^{-1}/m^3) of a resource species is calculated as the numerical abundance (NA_i) of the resource species times the productivity of an individual, approximated by $BM_i^{3/4}$. The available resource biomass B_j and the available resource productivity Π_j both require trivariate information regarding the food web (the resource species of each consumer, and the consumer species of each of those resource species), body masses, and numerical abundance. In these measures, dividing by the number of consumer species V_i reflects the crude assumption, made for want of better information, that each consumer of a given resource species gets an equal share of the resource's biomass or productivity. This crude assumption could be refined if quantitative data were available on the flows of energy along each trophic link. A random variable, its frequency distribution, or a set of numbers is said to be right-skewed if its third central moment is positive, left-skewed if its third central moment is negative, and symmetric if its third central moment is zero. (The third central moment is the sum of the cubes of the deviations of each number from the mean.) A random variable is said to be right-log skewed if the logarithm of the random variable is right-skewed.

Departure from normality of a distribution is assessed using measures of kurtosis and symmetry (D'Agostino and Pearson, 1973). Characteristics of the observed food web are compared with predictions of a null-model. An appropriate null-model for the trophic-species web is the cascade model (see Section III.A.1). The cascade model's predictions for the mean and expected maximal food chain length, number of basal, intermediate, and top species, and number of links among these species categories were calculated using the formulas in Cohen *et al.* (1986). All logarithms in this chapter are calculated with base 10.

III. THEORY: INTEGRATING THE FOOD WEB AND THE DISTRIBUTIONS OF BODY SIZE AND ABUNDANCE

This section outlines quantitative models and qualitative theoretical arguments to guide the analysis of the data in subsequent sections.

A basic question of community ecology is whether "the populations at a site consist of all those that happened to arrive there, or of only a special subset, those with properties allowing their coexistence" (Elton, 1933). Many ecologists probably agree that communities are not purely randomly constituted, apart from stochastic processes (e.g., those related to colonization and extinction, MacArthur and Wilson, 1967). For example, it is well known that large species usually are less numerically abundant and are positioned higher in a food web than small species.

Our goal is to shed additional light on the structure of an ecological community by looking in detail at the univariate, bivariate, and trivariate patterns that involve the food web and the distributions of body size and abundance in a community (Table 1). This theoretical section reviews some simple models of these patterns. The models use only a few input variables. The models will be tested in section V using the data described in Section IV.

A. Predicting Community Patterns

1. The Cascade Model

The cascade model of food web structure tries to predict multiple food web properties from the simplest assumptions possible. A leisurely nontechnical summary of the cascade model and its motivation is given by Cohen (1989). Cohen *et al.* (1990) give a detailed theoretical and empirical exposition. Carpenter and Kitchell (1993a) also use the term "cascade model." Their model describes the dynamics of multiple populations interacting through food webs following major perturbations. As an example of a "trophic cascade" in Carpenter's sense, an increase in the abundance of the top trophic level leads to alternating decrease and increase in the abundance of trophic levels below. In this chapter *cascade* refers only to the following strictly static model of food web structure in the sense of Cohen *et al.* (1990).

Let S denote the number of trophic species in a community. Suppose the trophic species can be ordered from 1 to S (although this ordering is not a priori visible to an observer), and suppose that the ordering specifies a pecking order for feeding, so that any species j in this hierarchy or cascade can feed on any species i only if $i < j$ (which doesn't necessarily mean that j does feed on i, only that j can feed on i). Thus, species j cannot feed on any species with a number k if $k \geq j$. Second, the cascade model assumes that each species eats any species below it according to this numbering with probability d/S, independently of all else in the web. Thus, the probability that species j does not eat species $i < j$ is $1 - d/S$. These assumptions—that the species are ordered and that the probability of feeding is proportional to $1/S$, and that different feeding links are present or absent independently of one another—are all there is to the cascade model.

The cascade model has one parameter, d. To compare the model with an individual food web, the parameter d may be estimated from the observed number of species S and the number of links L as $d = 2L/(S - 1)$. To compare the model with the properties of a collection of food webs, assuming that the parameter d is the same in all of them, the parameter d may be estimated from the total number of species and the total number of links in all webs combined or from the set of pairs (S, L) for each web. All predictions derive solely from the number of species and the number of links. No other parameters are free.

The cascade model makes a surprising variety of predictions about food webs (Cohen, 1989; Cohen *et al.*, 1990, 1991) such as the number of basal, intermediate, and top species; the number of food chains; the mean food chain length; the maximum food chain length; and the numbers of basal-intermediate, basal-top, intermediate-intermediate, and intermediate-top links. These predictions are computed and compared with observations, shown later in Table 3. The cascade model also predicts, for example, how the maximum chain length should change as the area from which a food web is sampled increases (Cohen and Newman, 1991), how the relative frequency of intervality among food webs should change with increasing numbers of trophic species (Cohen and Palka, 1990), and how various proportions of links and species should change with increasing numbers of trophic species. The cascade model predicts that the mean vulnerability of a species should increase linearly as trophic position goes from high (top predators) to low (primary producers) within a community, providing theoretical support for a prediction of Menge and Sutherland (1976). Predictions of the cascade model are not always confirmed. Several elaborations of the cascade model have been proposed (e.g., Cohen, 1990, 1991; Cohen *et al.*, 1993; Solow and Beet, 1998; Williams and Martinez, 2000).

The diversity of predictions from the cascade model and some of its elaborations is important, not because the predictions are always correct (they are not), but because so many different aspects of food webs derive from a few simple assumptions, and are therefore not independent of one another. The important message of the cascade model and of sufficiently analyzed related food web models is that superficially diverse aspects of food webs vary in coordinated ways as a result of simple underlying mechanisms.

It is worthwhile to present and discuss the cascade model even if it sometimes makes predictions that are inconsistent with observations. Caswell (1988) argued persuasively that "models are to theoretical problems as experiments are to empirical problems." In particular, the failure of a model to reproduce some empirical observations or patterns may be a source of insight, stimulating further thought and eventually further theoretical understanding. The cascade model does not claim that real food webs are constructed as described by the cascade model, only that such a simple set of assumptions is capable of integrating in a single perspective a large variety of observable aspects of single food webs and collections of food webs (Caswell, 1988). Kenny and Loehle (1991) make a similar claim for their "random web" model, a model that is biologically even more rudimentary than the cascade model. The cascade model made possible new ways of thinking about the properties of ensembles of food webs, demonstrated the conceptual linkage among these properties, and continues to provide a baseline against which variations in individual food webs can usefully be evaluated, as in this chapter. The deviations between at least some of the food web statistics computed for the Tuesday Lake data and the predictions of the simple cascade model are interesting since they pose the challenge of identifying the biological mechanisms at work that are ignored by the model.

2. Body Mass Rank and the Distributions of Body Mass, Abundance, and Trophic Height

Cohen (1991) hypothesized that, on average, the body masses of the species in a community could be related to their rank in body size. Two simple alternatives are that body mass (BM) is related to body size rank i starting from the largest species either allometrically

$$BM_i = \alpha i^\beta$$

or exponentially

$$BM_i = \alpha \beta^i$$

where α and β are two constants (with unknown values for the moment). If either equation is approximately true and if α and β are known, then the distribution of body size in a community can be predicted from the number of species. This relationship could be used to predict the numerical abundance (NA) of the species. Assuming that

$$BM_i = \alpha i^\beta$$

and that numerical abundance is allometrically related to body mass by

$$NA_i = \gamma(BM_i)^\delta$$

as has often been found (Damuth, 1981; Peters and Wassenberg, 1983; Blackburn and Gaston, 1999), then

$$NA_i = \gamma(BM_i)^\delta = \gamma(\alpha i^\beta)^\delta = \gamma \alpha^\delta i^{\beta\delta} = \varepsilon i^\varphi$$

That is, numerical abundance is allometrically related to body size rank i. Alternatively, if

$$BM_i = \alpha \beta^i$$

and

$$NA_i = \gamma(BM_i)^\delta$$

then

$$NA_i = \gamma(BM_i)^\delta = \gamma(\alpha\beta^i)^\delta = \gamma \alpha^\delta \beta^{\delta i} = \varepsilon \omega^i$$

meaning that numerical abundance is exponentially related to body size rank i.

If body mass and numerical abundance are allometrically related to body mass rank, then so is biomass abundance, with an exponent that is determined by the exponents for body mass and numerical abundance. Finally, if larger species on average are found higher up in a food web than small species, the trophic height of species could potentially be related to the rank (or log rank) in body mass. In principle, if the simple models presented here can be validated, the body mass, abundance, and trophic height of the species could be predicted using only the number of species and a few input parameters (the regression coefficients). To be practically useful, however, the regression coefficients of the relationships must be known. As with the cascade model, these simple relationships could provide a baseline against which observations in real communities can be compared, for example, to identify groups of species within a community that deviate from a predicted relationship or communities that behave differently (e.g., because they have been disturbed). If these relationships held in Tuesday Lake prior to the 1985 intervention, as we shall see, then we may hypothesize that body size, abundance, and trophic height in the perturbed community of Tuesday Lake in 1986 will be less predicted by the rank in body size than in 1984.

By treating the number of species as the independent variable to predict the distributions of body size, abundance, and trophic height, we do not mean to suggest that body size is independent of for example trophic organization, or that the number of species in a community is prior to and independent of the distributions of body size, abundance, and trophic organization. We are for the moment interested in analyzing how far this extremely simple approach, free of biological mechanisms, can go.

The relationships described above aim to predict only the expected value (body mass, abundance, or trophic height) of a species and neglect all variation in the dependent variable. Since more and more confounding factors may be included as the rank in body mass is used to predict successively the body mass, numerical abundance, and finally biomass abundance, we predict (not surprisingly) that the distribution of body mass will be best predicted by the rank in body mass, followed by numerical abundance, and then by biomass abundance.

B. The Distribution of Body Sizes

The body size of an organism matters ecologically and evolutionarily, and so does the ensemble of body sizes in an ecological community. Many ecological traits (e.g., generation time, clutch size, ingestion rate, and population density) are significantly correlated with body size (Peters, 1983; Calder, 1984). Harvey and Purvis (1999) point out that some recent mathematical models (Charnov, 1993; Kozlowski and Weiner, 1997) suggest that an organism's body size is an adaptation to its life history characteristics rather than the other way around. Notwithstanding these models, it seems likely that body size and other life history characteristics are jointly determined. Large differences in body size (and thus also in demographic rates) between the species in a community can lead to dynamics on several time scales (e.g., Kerfoot and DeAngelis, 1989; Muratori and Rinaldi, 1992). The ratio of the turnover rates of the primary producers and consumers, as a function of their relative sizes, may affect the stability of the system. Conversely, constraints imposed by requirements for stability could affect the distribution of body size. Body size has also been shown to affect extinction risks of carnivores and primates (Purvis *et al.*, 2000). On an ecological time scale, the feeding interactions of animals are probably constrained by body size, but on an evolutionary time scale, feeding interactions may affect body size.

In community ecology, much attention has been devoted to the shape of body size distributions and how they are affected by sampling biases and spatial scale (see e.g., Brown and Nicoletto, 1991; Blackburn and Gaston, 1994). Histograms of the number of species in logarithmic body size classes are typically right skewed. In global assemblages or for single taxa such as birds, mammals, or fish, the suggested slope of the right tail on log-log scales is −2/3 for body mass (May, 1986) but varies considerably among many studies (Loder *et al.*, 1997). Few studies of the body size distribution in entire community assemblages are available. Holling (1992) proposed that a few key biotic and abiotic processes in ecosystems may be responsible for generating spatial and temporal structure, and that the discontinuity in space or time of these processes leads to clumps and gaps in the distribution of body sizes in communities. If Holling's hypothesis holds generally, the distribution of body sizes in Tuesday Lake should show clumps and gaps (Havlicek and Carpenter, 2001).

By definition, the species in a guild or taxonomic group all have similar (but not identical) niches or trophic positions. Assuming that one body size (or body size class) is best adapted to the particular way of living of the guild, this size class can be expected to have more species than other size classes. In a community with many different guilds, the trophic positions and body sizes could be expected to vary much more than within taxonomic groups. Here, other mechanisms such as speciation, immigration, and extinction rates relative to body size may be important in shaping the body size distribution. Thus, for a community, a right-log skewed, perhaps even log-hyperbolic, distribution may be more likely. (The hyperbolic distribution has log-linear tails. The body size distribution is log-hyperbolic if the logarithm of body size is hyperbolically distributed.)

The shape of the size distribution of species may change with the geographical range of the investigation (Blackburn and Gaston, 1994). If the body size and the geographical range of species are positively correlated (as suggested by Brown and Maurer, 1987 and supported by empirical data), then as the geographical range of a study increases, relatively more small-bodied than large-bodied species will be added to the distribution, because a large fraction of the large species will be found at small scales but only a limited fraction of the small species. Local communities would then be expected to have a shallower slope of the right tail of the relationship between body size and number of species than regional assemblages. The slope of regional relations should in turn be shallower than a global relationship. Based on these arguments, we hypothesize that the slope of the right tail of the body size distribution in Tuesday Lake will be less negative than $-2/3$.

C. Rank-Abundance and Food Web Geometry

The rank-abundance relationship has been studied principally in competitive communities, guilds (functional groups), or taxonomical groups, with a focus on organisms thought to compete for some limiting resource(s) in an ecological community. Early work (MacArthur, 1957, 1960; Cohen, 1966) discussed the effect of various resource partitioning mechanisms among organisms on the distribution of abundance.

Here we display the rank-abundance relationship across all the recorded species in Tuesday Lake, including primary producers, primary and secondary consumers, and several guilds and taxonomical groups. Our hypothesis is that the rank-abundance relationship across all species is affected by the shape of the food web. Just as past analyses of the rank-abundance relationship have been carried out to shed light on the resource partitioning mechanism in a particular group of species, the rank-abundance relationship of a community-wide food web reflects and can shed light on the geometric shape of the food web and energy flows through the community.

Assuming (as we will demonstrate later) that body size generally increases and numerical abundance generally decreases from the bottom (primary producers) to the top (top predators) of a food web, a pyramidal web (wide base and narrow top) implies a large number of small and numerically abundant species and fewer large and relatively rare species. In comparison with a pyramidal web, a more rectangular food web would have relatively fewer small and numerically abundant species. If numerical abundance decreases exponentially with every step in a food chain, so that the numerical abundance of a predator on average is a constant small fraction of the numerical abundance of its prey, then the numerical abundances of the species in a simple food chain would follow a geometric series (i.e., a linear decrease in log numerical abundance as a function of abundance rank). Extending this line of reasoning to a whole food web suggests that the shape of the rank-log abundance relationship may reflect the shape of the food web. For example, a pyramidal shape of the food web (plus an exponential decrease in numerical abundance with trophic height) could imply a concave rank-log abundance relationship. We predict that the geometric shape of the food web of Tuesday Lake will be reflected in the rank-log abundance relationship, and conversely that the geometric shape of the food web of Tuesday Lake can be anticipated from the shape of the rank-log abundance relationship.

D. Linking the Food Web to the Relationship Between Body Size and Numerical Abundance

Studies of the relationship between body size and numerical abundance in animals have, with a few exceptions (Marquet et al., 1990; Cyr et al., 1997a), concentrated on "regional" or "global" collections of species (Mohr, 1940; Damuth, 1981; Peters and Wassenberg, 1983; Peters and Raelson, 1984; Brown and Maurer, 1987) or particular taxa or functional groups within local communities (e.g., Morse et al., 1988). Most studies showed that log numerical abundance decreases linearly as log body size increases. The slopes of log numerical abundance as a function of log body size relationship have been more negative at regional than at local scales (Blackburn and Gaston, 1997; Enquist et al., 1998). More restricted taxonomic groups have a less negative slope than broader aggregations (Peters and Wassenberg, 1983; Cyr et al., 1997b). Some investigations, however, claimed that the relationship is polygonal (Brown and Maurer, 1987; Morse et al., 1988) or otherwise nonlinear (Silva and Downing, 1995). Blackburn and Gaston (1997) reviewed different forms of the abundance-body size relationship.

Blackburn and Gaston (1999) also reviewed mechanisms proposed to explain the abundance-body size relationship, including the "energetic constraint mechanism." This hypothesis asserts that the slope of the relationship is a function of the basal metabolic rate of organisms and the amount of energy used by populations. Other explanations for the observed relationship between numerical abundance and body size include sampling from the distributions of abundance and body size (the "concatenation mechanism," Blackburn et al., 1993) or body-size–related extinction risks (the "differential extinction mechanism"). Blackburn and Gaston (1999) concluded that no single mechanism adequately explains the published abundance-body size relationships.

Since metabolic rate scales as BM^α with α claimed to be $\frac{3}{4}$ (Kleiber, 1932; Hemmingsen, 1960) or $\frac{2}{3}$ (Heusner, 1982; Dodds et al., 2001), metabolic efficiency should have a significant effect on numerical abundance at least over large ranges in body size, that is, across all species in a community.

Across restricted ranges in body size, other mechanisms such as interspecific interactions could overshadow the effect of metabolic efficiency. Here, the potential effect of the trophic structure of a community on species' resource uses and abundance-body size relationships will be explored.

Assume that the numerical abundance of a consumer population i is approximately proportional to the total amount of resources available to the consumer population per unit time (i.e., resource supply rate, ρ_i) divided by the resource use per consumer individual per unit time, and that the resource use per individual per unit time is proportional to the metabolic rate of individuals (MR). The metabolic rate of individuals is allometrically related to body mass as

$$MR \propto BM^\alpha$$

where $\alpha < 1$ and α is often claimed to approximate $\frac{3}{4}$ (e.g., Hemmingsen, 1960). Symbolically,

$$NA_i \propto \rho_i/MR_i \propto \rho_i \times BM_i^{-\alpha}$$

(Carbone and Gittleman, 2002). If each consumer species on average has the same amount of net resources available to it (i.e., $\rho_i = c$, a totally arbitrary assumption which we shall challenge in the next paragraph), then

$$NA \propto BM^{-\alpha}$$

so that the slope $-\alpha$ of log numerical abundance as a function of log body mass for consumer species is the negative of the allometric exponent α relating body mass to metabolic rate (see Enquist et al., 1998).

However, the resource supply rate is probably not the same for all consumer species in a community. The structure of the food web, the positions of species within it, and the efficiency with which species extract resources will affect species' resource supply rates. Consequently, the resource supply rate could increase or decrease with increasing consumer body size or trophic height of a consumer species. The larger a species is, the more available prey species there are. On the other hand, prey species are in general shared by other consumers, so the larger a species is, the higher in the food web it may feed, with possibly less energy available due to ecological efficiencies. Unless larger species are more omnivorous than smaller species, the amount of resources available to a larger species could decrease.

If consumer numerical abundance can be divided by an estimate of the resource supply rate to each consumer, theory suggests (Carbone and Gittleman, 2002) that the slope should be closer to $\frac{-3}{4}$ since

$$NA_i \propto \rho_i \times BM_i^{-\alpha} \iff NA_i/\rho_i \propto BM_i^{-\alpha}$$

We predict that if the slope of log numerical abundance as a function of log body mass in Tuesday Lake deviates from $\frac{-3}{4}$ on log-log scales, then the slope of numerical abundance of consumers divided by an estimate of the resource productivity available to each consumer, versus the body mass of the consumer, will be closer to $\frac{-3}{4}$ on log-log scales. To infer more specifically if resources available to a consumer change with body size or trophic position would require species-specific data on energy flow. In the absence of such data, we analyze these relationships indirectly.

If the resource supply rate is the same for all consumer species and if consumers' metabolic rates are allometrically related to body mass by an exponent of $\frac{3}{4}$, then each consumer species should be found along a line with slope $\frac{-3}{4}$ in the log body mass-log numerical abundance plane, starting from the resource species (i.e., the point [log BM_{prey}, log NA_{prey}]). Deviations from this prediction for individual pairs of consumer and resource species could indicate either that the prey species has more than one predator species, or that the predator species has more than one prey species. The former means that the productivity of a particular prey must be shared with several predator species, leading to lower than expected numerical abundance of the consumer species (thus making the slope steeper than $\frac{-3}{4}$). The second case means that a particular consumer species has more than one prey species to provide resources, leading to a higher than expected numerical abundance of the consumer species (thus making the slope less steep than $\frac{-3}{4}$). Based on these arguments, we predict that in Tuesday Lake, for individual pairs of consumer and resource species, there will be: (1) a positive relationship between the slope of log numerical abundance as a function of log body mass on the one hand, and the consumer's trophic generality on the other; and (2) a negative relationship between the slope of log numerical abundance as a function of log body mass on the one hand, and the prey species' trophic vulnerabilities on the other.

E. Trophic Pyramids and the Relationship Between Consumer and Resource Abundance Across Trophic Levels

In many ecosystems, predators are larger and less numerically abundant than their prey, if parasites are ignored (Darwin and Wallace, 1858). Elton (1927, p. 69) noted that, "... animals at the base of a food chain are relatively abundant, while those at the end are relatively few in numbers ...." A bar plot of numerical abundance in different trophic levels in a community (later illustrated by Fig. 8A, B) often produces the well-known "pyramid of numbers," a monotonic decrease in numerical abundance with increasing trophic level. Alternatively, a bar plot of biomass in different trophic levels may give a "pyramid of biomass" (see Odum, 1983; Wetzel, 1983) or an inverted pyramid (illustrated later by our Fig. 8C, D).

The standing crop of biomass may increase or decrease with increasing trophic height, depending on the balance between the loss of energy along each link in a food chain and the residence time of the energy in the individuals at successive trophic levels or nodes in a food chain. The reciprocal of the residence time is defined as the *turnover rate*. Energy is lost at every trophic transfer in a food chain, but if the resource has a much faster turnover rate than the consumer, a loss in the transfer of energy to the consumer may be compensated for by a longer residence time of energy in the consumer, allowing the standing crop of a consumer to equal or exceed that of its resource (e.g., Harvey, 1950).

Elton (1927) referred to a pyramid of numbers only. Many textbooks use "trophic" or "ecological" pyramids more broadly to refer to the pattern of numerical or biomass abundance, or productivity in successive trophic levels. The very concept of trophic levels has been criticized as an excessive simplification of the trophic structure of communities (e.g., Cousins, 1987). In estimates of the numerical or biomass abundance at different trophic levels in ecosystems, "trophic levels" often are specified as primary producers, primary consumers (herbivores), and secondary consumers (carnivores). Studies with estimates of autotrophic and heterotrophic biomass in freshwater plankton and marine ecosystems are reviewed by Del Giorgio and Gasol (1995) and Gasol *et al.* (1997), respectively. Baird and Ulanowicz (1989) reported energy flows of the Chesapeake Bay food web and of an aggregated food chain with eight trophic levels.

The change in abundance across trophic levels depends in part on the relationship between consumer and resource abundance at the species level. Here we consider a consumer c that feeds on a single resource r. The ratio between consumer and resource numerical abundance in a community may be modeled by using the same energetic assumptions as above (Section III.D) to relate the numerical abundance NA_c of consumer c to its body mass BM_c and to the productivity ρ_r of resource r. We make the assumption, plausible for this situation, that the abundance of the resource depends on its own productivity while the abundance of the consumer depends on the food it can sustainably extract from its resource, which is proportional to the resource productivity. If

$$NA_c \propto \rho_r \times BM_c^{-\alpha} \iff NA_c/\rho_r \propto BM_c^{-\alpha}$$

and

$$\rho_r \propto BM_r^{\alpha} \times NA_r$$

then

$$\frac{NA_c}{NA_r} \propto \left(\frac{BM_r}{BM_c}\right)^{\alpha} \text{ and } \frac{BA_c}{BA_r} \propto \left(\frac{BM_c}{BM_r}\right)^{1-\alpha}$$

The ratio of consumer to resource numerical abundance is predicted to be proportional to the consumer-resource body size ratio raised to the power α. The ratio of consumer to resource biomass abundance is predicted to be proportional to the consumer-resource body size ratio raised to the power $1-\alpha$. Consequently, the larger in size a predator is relative to its prey, the smaller the ratio between predator and prey numerical abundance is predicted to be, but the larger the ratio between predator and prey biomass abundance is predicted to be. This illustrates the well-known fact that even though predator biomass abundance often tends to be smaller than prey biomass (at least in terrestrial systems), a biomass abundance ratio greater than unity is possible if the difference in turnover rates of the predator and prey is large enough. Because of the allometric relation between turnover rates and body size, a biomass abundance ratio can exceed unity if the consumer is much larger than the resource. We predict that in Tuesday Lake: (1) the ratio of predator to prey numerical abundance will be positively correlated to the prey-predator body mass ratio and (2) the ratio of predator to prey biomass abundance will be positively correlated to the predator-prey body mass ratio. We also predict that the slope of the former relationship should be greater than that of the latter.

These predictions, which apply directly only to a pair of species consisting of one prey and one predator or one resource and one consumer, also have implications for food chains and food webs. If the predator-prey body mass ratio remains constant within a food chain, the ratio of predator to prey biomass abundance is predicted not to change systematically along the food chain. In a food web of cross-linked food chains, the picture could be more complicated. By analogy with the predictions for food chains, we predict that the changes in biomass and numerical abundance across trophic levels in Tuesday Lake will correlate with the average ratios in body mass between the species on different trophic levels. A small change in average trophic level body mass is predicted to be associated with a small change in trophic level numerical abundance between two trophic levels and with a decrease in trophic level biomass abundance. A large change in average trophic level body mass is predicted to be associated with a large change in numerical abundance between two trophic levels and possibly with an increase in biomass abundance from one trophic level to the next.

IV. DATA: TUESDAY LAKE

Tuesday Lake is a small, mildly acidic lake in Michigan (location 89°32′ W, 46°13′ N). Carpenter and Kitchell (1993b) described the physical and chemical characteristics of the lake. Summers are cool and winters are cold. Ice covers the lake from November to late April, on average, and oxygen is depleted during most winters. The fish populations are unexploited and the drainage basin undeveloped. For most of the lake's history, the fish fauna has been typical of winterkill lakes of the region. In 1984, the fish were three (mainly zooplanktivorous) species: 90% northern redbelly dace (*Phoxinus eos*), 5% finescale dace (*Phoxinus neogaeus*) and 5% central mudminnow (*Umbra limi*) (Hodgson *et al.*, 1993). The dace are zooplanktivores capable of altering the size and species composition of a zooplankton community. Since 1984, Tuesday Lake has been part of a series of whole-lake experiments conducted by S. R. Carpenter and colleagues (summarized in Carpenter and Kitchell, 1993a). Data from 1984 and 1986 are analyzed here.

A. The Manipulation

Prior to 1985, Tuesday Lake lacked naturally occurring large piscivores. The first experiment by Carpenter and colleagues consisted of removing 90% of the fish biomass in May and July of 1985 and replacing the planktivorous species with one species of largely piscivorous fish, largemouth bass (*Micropterus salmoides*) from a nearby lake (Table 2). Largemouth bass is a potential keystone predator (Hodgson *et al.*, 1993).

Bass consumed practically all the remaining dace shortly after the introduction. The survival rate of the bass was high and the population recruited successfully in both 1985 and 1986 (Hodgson *et al.*, 1993). However, few members of the cohort of 1985 survived through the winter of 1985–86 (due to a combination of predation by adult bass and size-selective winter mortality), so small juvenile largemouth bass can be considered absent throughout 1986.

The effects of the manipulation were documented by Carpenter and Kitchell (1988, 1993a). Bass introduction in Tuesday Lake caused a dramatic

Table 2 The manipulation of the fishes of Tuesday Lake in 1985

Date	Number of individuals removed	Number of individuals added
May 1985	39,654 redbelly dace (*Phoxinus eos*) 2,692 finescale dace (*Phoxinus neogaeus*) 2,655 mudminnows (*Umbra limi*)	375 largemouth bass (*Micropterus salmoides*, 47.5 kg)
July 1985	None	91 largemouth bass (*Micropterus salmoides*, 10.1 kg)
Total	45,001 individuals (56.4 kg)	466 individuals (57.6 kg)

reduction in vertebrate zooplanktivory. Consequently, the zooplankton assemblage shifted from dominance by small-bodied species (e.g., *Bosmina*, rotifers, and small copepods) to dominance by large-bodied cladocerans (i.e., *Daphnia*), along with a substantial decrease in chlorophyll concentrations and primary production. These changes are examples of "trophic cascades" in the sense of Carpenter (Carpenter *et al.*, 1985; Carpenter and Kitchell, 1993a).

B. The Data

To establish the pelagic food web of Tuesday Lake, intensive diet data were collected for the fish and *Chaoborus*. Stomach content analyses were done on 434 largemouth bass from Tuesday Lake. Minnow and dace diets were based on analyses in 1984 of 40 individuals of each species (Cochran *et al.*, 1988). *Chaoborus* diets were measured by Elser *et al.* (1987a). For predaceous crustaceans, we judged diets on the basis of personal communications with S.I. Dodson and T.M. Frost. For herbivorous zooplankton, grazing experiments conducted in Tuesday Lake or nearby lakes were used whenever possible (Bergquist, 1985; Bergquist *et al.*, 1985; Bergquist and Carpenter, 1986; Elser *et al.*, 1986, 1987b; St. Amand, 1990). Most herbivorous zooplankton are filter-feeders and the filtering apparatus sets limits on the sizes of phytoplankton they can extract. Judgments on the grazer-phytoplankton links considered whether the resource was sufficiently small and vulnerable and co-occurred with the consumer. These decisions may confound the relationships between body size and trophic structure. Diets of some predators change extensively over ontogeny, with consequences for "trophic cascades" (Carpenter *et al.*, 1985). Diets reported here are for the body sizes and life stages present in the lake in either 1984 or 1986. If multiple life stages or a range of body sizes were present, the data represent the aggregate diet for the species during the time period. No information on parasites of the pelagic species of Tuesday Lake is available, and no information on the pelagic microbial community is included. Table 3 summarizes statistics of the food web.

Physical and chemical variables and plankton abundance (not *Chaoborus*) were censused weekly from May to September (Carpenter and Kitchell, 1993b). Night tows and minnow trappings were used every two weeks to census the abundance of *Chaoborus* and planktivorous fish respectively. Largemouth bass were censused twice a year (at the beginning and end of the field season) by angling and electrofishing. The primary data on Tuesday Lake included the average body length (m) of the species, individual volume (m^3) of the phytoplankton, body mass (kg) of the zooplankton (including *Chaoborus*) and fish, and numerical abundance (individuals/m^3). These data were then converted to uniform measures for all species and combined

Table 3 Statistics for the unlumped and trophic species webs of Tuesday Lake

Statistic	Unlumped web, 1984	Trophic web, 1984	Unlumped web, 1986	Trophic web, 1986
Species	56 (50)[a]	27 (21)[a]	57 (51)[a]	26 (20)[a]
Phytoplankton species	31 (14)[b]		35 (18)[b]	
Zooplankton species	22 (6)[b]		21 (5)[b]	
Fish species	3 (3)[b]		1 (1)[b]	
Basal species[a]	25	8 (3.1)[c]	29	6 (3.6)[c]
Intermediate species	24	12 (14.7)[c]	20	12 (12.8)[c]
Top species	1	1 (3.1)[c]	2	2 (3.6)[c]
Food chains	4836	214 (263)[c]	885	59 (115)[c]
Mean food chain length[d]	4.64	3.68 (5.08)[c]	4.21	3.47 (4.30)[c]
Maximum food chain length[d]	7	6 (10)[c,e]	6	5 (8)[c,e]
Links	269 (264)[f]	71 (67)[f]	241 (236)[f]	56 (52)[f]
Basal-intermediate links	166	31 (14.74)[c]	158	20 (12.75)[c]
Basal-top links	0	0 (3.12)[c]	7	2 (3.6)[c]
Intermediate-intermediate links	87	27 (34.39)[c]	68	27 (22.9)[c]
Intermediate-top links	11	9 (14.74)[c]	3	3 (12.75)[c]
Connectance[a,f]	0.2155	0.3190	0.1851	0.2737
Consumers per resource species[a,f]	5.39	3.35	4.82	2.89
Resources per consumer species[a,f]	10.56	5.15	10.73	3.71
Consumers per phytoplankton species[a,f]	5.35		4.71	
Consumers per zooplankton species[a,f]	4.36		3.38	
Resources per zooplankton species[a,f]	10.68		11.10	
Resources per fish species[a,f]	9.67		3	

[a]Isolated species excluded.
[b]Number of unique species in parenthesis (i.e. species that occurred in that year only).
[c]Numbers in parenthesis indicate cascade model predictions.
[d]Number of links.
[e]Longest food chain with an expected frequency greater than one.
[f]Cannibalistic links excluded.

with the trophic data (see above) so that the data analyzed here (Appendices 1 and 2 for 1984 and 1986, respectively) consist of: (1) a list of species; (2) the predators and prey of each species; (3) the trophic height of each species; (4) the average body mass (kg) of the species; (5) numerical abundance (individuals/m^3) of the species; and (6) the biomass abundance (kg/m^3) of the species, which is the product of the body mass times numerical abundance. The data represent seasonal averages during summer stratification.

For the plankton and planktivorous fish, the concentrations of individuals are the average over the weeks in which the taxon was present. For the piscivorous fish, numerical abundance was calculated as the average of the censuses at the beginning and end of the field season. For the phytoplankton and small zooplankton (<0.5 mm), numerical abundance at each census was determined by counting individuals on a lattice (a microscope slide marked with a rectangular grid to minimize confusion while counting) until the standard error of the mean number of individuals per subsample was less than 10%. For small phytoplankton, a minimum of 10 microscope fields at a magnification of 400× were counted (minimum of 100 individuals). For larger phytoplankton, a minimum of 15 fields at a magnification of 200× were counted (minimum of 300 individuals). For larger zooplankton (>0.5 mm diameter), the entire sample was counted.

Body size (length, mass, or volume) was obtained by measuring individuals from Tuesday Lake, in general, until the standard error of the mean was less than 10%. The values are reported as average values. For species with highly variable size, such as colonial species, the range and geometric mean were also reported. For the phytoplankton, the size data are for "algal units:" single cells were measured for solitary species and the size of the colony was measured for colonial species. The only colonial zooplankton in the data is *Conochilus* sp. in 1986, for which colony size is reported. From individual volume (and an assumed density of 1 kg/l), the body mass (kg) of the phytoplankton was estimated. For all species, body mass is kg fresh weight. Fresh weights include a variable proportion of water and cannot be converted to elemental compositions without additional information or assumptions.

The numerical abundance of all species used for all calculations here is the number of individuals per cubic meter in the water volume where the consumers feed (i.e., in the epilimnion), which is roughly equal to the photic zone in Tuesday Lake. Phytoplankton were sampled in the epilimnion. Zooplankton were sampled (using vertical net hauls during day time) over a water mass that is about six times the volume sampled for phytoplankton. Thus, the volume where zooplankton *live* is about six times the epilimnion volume where zooplankton *feed* on phytoplankton. The total sizes of the fish populations in the lake were estimated using mark-recapture and were then divided by the volume of the epilimnion. When zooplankton feed in the epilimnion (typically at night), their concentrations in this zone are considerably higher (Dini *et al.*, 1993) than in the volume where they live.

In Appendices 1A and 2A, in the columns headed *NA* (for "numerical abundance"), the values for phytoplankton and fish were used in statistical analyses without change, but the numerical abundance values for all zooplankton species in Appendices 1A and 2A were multiplied by 6 before use in the statistical analyses reported here. For example, if the stated values in Appendix 1A (for 1984) in the columns headed *BM* and *NA* were used without this adjustment by a factor of 6 for zooplankton numerical abundance, then the coefficients of the linear model

$$\log_{10}(NA) = a + b \times \log_{10}(BM)$$

considered in the first line of Table 6 for all species ("Total") would be $a = -3.6109$, $b = -0.8877$. However, because all the zooplankton numerical abundances were multiplied by 6 to convert all species to the same reference volume (namely, the epilimnion), the same regression analysis reproduces the values shown in the first line of Table 6 (namely, $a = -2.6863$, $b = -0.8271$). Section VII addresses the effect of multiplying zooplankton numerical abundances by 6.

V. RESULTS: PATTERNS AND RELATIONSHIPS IN THE PELAGIC COMMUNITY OF TUESDAY LAKE

This section can be regarded as an illustrated list (Table 1) of different ways to describe a community, using data on the body size, numerical abundance, and food web in the pelagic community of Tuesday Lake. Many of the relationships presented are previously well studied. To examine the effect of scale, we will compare data from the whole Tuesday Lake pelagic community with previously analyzed data on restricted taxa or data aggregated over several communities.

The three dimensions of our analysis are: the food web, body size, and species abundance. We start by looking at three-dimensional data, then move on to two-dimensional relationships followed by one-dimensional relationships. The three-dimensional perspective is the principal novelty this article offers. It permits ecologists to view Tuesday Lake in a series of new, coherent pictures and provides the baseline against which the two- and one-dimensional relationships will be compared.

A. Trivariate Distributions: Food Web, Body Size, and Abundance

In Tuesday Lake, small-bodied, numerically abundant species occur at low trophic heights, whereas larger-bodied and less abundant species occur at higher trophic heights (Figs. 1A, B, 2A, and B). Biomass abundance does not vary systematically with body mass or trophic height (Figs. 1A, B, 2C, and D).

In Fig. 2, phytoplankton, zooplankton, and fish form three distinct clusters. The data points plotting numerical abundance, body mass, and trophic height lie roughly on a diagonal between the lower left corner and the upper

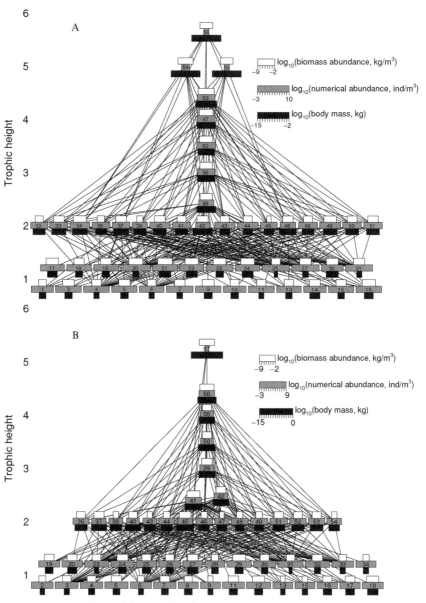

Figure 1   The unlumped food webs of Tuesday Lake in (A) 1984 and (B) 1986. The width of the black, grey and white horizontal bars shows the $\log_{10}$ body mass (kg), numerical abundance (individuals/m^3 in the epilimnion where species eat), and biomass abundance (kg/m^3 in the epilimnion where species eat), respectively, of each species. Species numbers refer to Appendices 1 and 2. The vertical positions of the species show trophic height (see text). Basal species have a trophic height of unity

right corner in Fig. 2A and B. Multiple regression yields trophic height = 0.3421 × log body mass −0.1040 × log numerical abundance +5.8697 (with squared multiple correlation coefficient $r^2 = 0.8404$). An interpretation of the first coefficient in this regression equation is that an increase in mean trophic height by one level is associated with an increase in body mass by a factor of more than 800 (because $1/0.3421 = 2.9231$ and $10^{2.9231} = 837.8$), if all else remains constant. However, in Tuesday Lake, an increase in body mass is usually closely associated with a decrease in numerical abundance. Variation in trophic height due to log body mass and log numerical abundance in combination can be attributed more to log body mass (controlling for log numerical abundance [the partial correlation coefficient of trophic height and log body mass, given log numerical abundance, is $0.5790, p < 0.01$]) than to log numerical abundance (controlling for log body mass [the partial correlation coefficient of trophic height and log numerical abundance, given log body mass, is $-0.2179, p > 0.05$]).

Among the phytoplankton in 1984, body mass and numerical abundance are negatively correlated, although all phytoplankton have a trophic height of 1 (Table 4). For 17 species of zooplankton with a trophic height of 2, body size and numerical abundance are significantly negatively correlated in 1984 ($r_{84} = -0.5262, p < 0.05$). In 1986, the negative correlation, although of similar magnitude, is not significant ($r_{86} = -0.4940, p > 0.05$).

Variations in numerical abundance are more closely associated with variations in body mass than with variations in trophic height. Bivariate correlations indicate that log body mass is more closely associated with log numerical abundance ($r^2 = 0.8414$) than trophic height is with log numerical abundance ($r^2 = 0.7628$) in Tuesday Lake in 1984 (Table 4). Multiple correlation analyses of log numerical abundance (dependent variable) on log body mass and trophic height (independent variables) show that trophic height adds little to explaining the variation in numerical abundance.

Figure 3, a new food web graph inspired by the diagrams in Cousins (1996) and Sterner et al. (1996), shows the food web of Tuesday Lake in the plane with abscissa log numerical abundance and with ordinate log body mass. Animal ecologists generally put log body mass on the abscissa, while plant ecologists generally put log body mass on the ordinate. Since food webs are conventionally represented with food flowing in an upward

---

by definition, but to allow for wider non-overlapping bars, the vertical positions of the basal species have been adjusted around unity. The horizontal position is arbitrary. Isolated species (see Appendices 1 and 2) are omitted. Species with a trophic height of unity are phytoplankton, those with a trophic height greater than 4.5 are fish, and those with intermediate trophic heights are zooplankton. Figure 1A is reprinted from Cohen et al. (2003) with permission from the National Academy of Sciences.

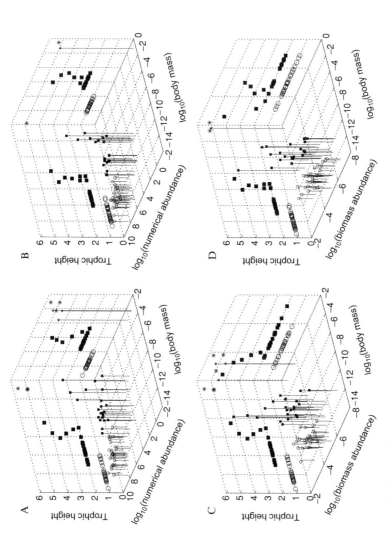

Figure 2 Body mass (kg), trophic height and abundance of the species in Tuesday Lake in 1984 (A & C) and 1986 (B & D). Numerical abundance (individuals/m^3 in the epilimnion where species eat) is shown in (A) and (B), and biomass abundance (kg/m^3 in the epilimnion where species eat) in (C) and (D). Circles = phytoplankton, squares = zooplankton, stars = fish. Small markers on

direction, we prefer the choice of axes customary among plant ecologists so that food usually flows upward (and from right to left) from smaller-bodied, more abundant prey to larger-bodied, rarer predators.

The slope of a trophic link that connects two species, a consumer and resource, in this diagram is defined as (log body mass of consumer − log body mass of resource) divided by (log numerical abundance of consumer-log numerical abundance of resource). The slope indicates the biomass ratio between a predator or consumer species and one of its prey or resource species. A slope of −1 indicates equal biomass abundance of predator and prey. A slope more negative (or less negative) than −1 indicates that the predator has greater (or smaller) biomass abundance, respectively, than the prey. The mean slope of all links that join individual pairs of consumers and resources was −1.1585 in 1984 and −0.8625 in 1986. The mean slope of all links does not equal the slope of the regression of log body mass as a function of log numerical abundance (Section V.B.2.a). Among all noncannibalistic trophic links, 62% in 1984 and 67% in 1986 connect a predator and a prey where the biomass abundance of the prey is smaller than that of its predator.

While the body mass of individual species increases almost 12 orders of magnitude and the numerical abundance of individual species decreases almost 10 orders of magnitude within the food web, biomass abundance increases on average two orders of magnitude from the bottom to the top of the food web (as expected: 12 − 10 = 2). Biomass abundance varies only five orders of magnitude over all species.

The food web diagram in Fig. 3 carries more information on the pattern of energy flow within a community than a traditional food web graph. We know of no other study that shows the joint variation in body size and numerical abundance, and thus in biomass abundance, of all the species in a community food web.

B. Bivariate Distributions

1. Food Web and Body Size

A data set that includes the food web and the body sizes of the species makes it possible to analyze the predator-prey body size allometry, body size versus trophic height as well as trophic generality and vulnerability versus body size (Table 1).

---

stems show the position of each species in the three-dimensional space. The base of the stems on the floor of the box and larger markers on the walls show the bivariate distribution of the species in two-dimensional spaces. Figure 2A and C is reprinted from Cohen *et al.* (2003) with permission from the National Academy of Sciences.

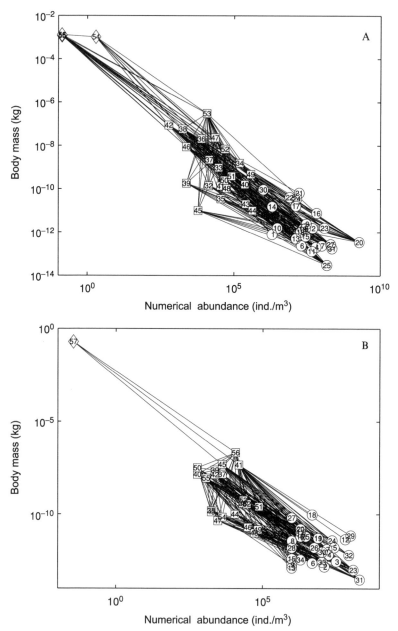

Figure 3   The Tuesday Lake food web in (A) 1984 and (B) 1986, plotted in the plane with abscissa measured by numerical abundance (individuals/m^3 in the epilimnion where species eat) and ordinate measured by body size (kg) on logarithmic scales for both axes. The center of a node locates the species identified by the number within

a. Predator-Prey Body Size Allometry

In aquatic and terrestrial habitats, if parasites and terrestrial herbivores are ignored, a predator is usually larger than its prey (Vézina, 1985; Warren and Lawton, 1987; Cohen *et al*., 1993) and predator size is in general positively correlated to prey size. In some systems, the ratio in body size of two predators that differ in body size is typically less than the ratio in body size of the prey of those predators, that is, comparatively speaking, prey body size increases faster than predator body size, so that the larger the prey is, the more similar its size is to the size of its predators (Vézina, 1985; Cohen *et al*., 1993).

These general patterns are predicted by the cascade model (Section III.A.I) if the ordering assumed in that model is interpreted as an ordering by body size. Under that interpretation, the cascade model assumes that predators are larger than their prey. The cascade model predicts that the larger the prey is, the more similar in size its average predator should be. The larger the predator, the less similar in size its average prey should be.

These relationships hold for Tuesday Lake (Fig. 4, Table 5). However, contrary to an assumption of the cascade model, the points that represent pairs of predator and prey in Fig. 4 are not randomly distributed above the diagonal. Rather, the data points lie in a wide band above the diagonal. The largest consumers (fish) do not eat the smallest prey (phytoplankton). Though Tuesday Lake conforms well to the predictions of the cascade model for the qualitative relationships between predator and prey size, the trophic links are not distributed, as the cascade model assumes, with equal probability between each predator and any species smaller than the predator.

In this deviation from the "equiprobability assumption" of the cascade model, Tuesday Lake is not alone. In 16 published food webs for which estimates of adult body masses were available, Neubert *et al*. (2000) found some evidence of departure from the equiprobability assumption in 7 of the 16 webs (at a significance level of $p \leq 0.06$). In six of these webs, the probability of a trophic link was affected by the identity of the predator species. This deviation from the original cascade model is captured in some generalizations (Cohen, 1990).

If species are sorted by their body size and isolated species are discarded, the resulting predation matrix has 269 nonzero entries (trophic links), of which 262 are from smaller prey to larger consumers. This finding must be interpreted cautiously because, as noted above, relative body sizes were

---

the node. Edges connect consumer species to the species they eat. Isolated species (see Appendices 1 and 2) are omitted. Circles = phytoplankton, squares = zooplankton, diamonds = fish.

Table 4 Correlations among body size, abundance and trophic height in Tuesday Lake[a]

Variables	Year	Phytoplankton			Zooplankton			Total		
		r	p	n	r	p	n	r	p	n
$\log_{10}(BM)$ vs. $\log_{10}(NA)$	1984	-0.5615	$p < 0.002$	31	-0.5366	$p < 0.02$	22	-0.9175	$p < 0.001$	56
	1986	-0.2723	$p > 0.1$	35	-0.4843	$p < 0.05$	21	-0.8665	$p < 0.001$	57
$\log_{10}(BM)$ vs. $\log_{10}(BA)$	1984	0.7028	$p < 0.001$	31	0.7982	$p < 0.001$	22	0.4343	$p < 0.001$	56
	1986	0.6885	$p < 0.001$	35	0.8740	$p < 0.001$	21	0.5212	$p < 0.001$	57
$\log_{10}(BM)$ vs. TH	1984	0			0.6398	$p < 0.002$	22	0.9135	$p < 0.001$	50[b]
	1986	0			0.6152	$p < 0.005$	21	0.8804	$p < 0.001$	51[b]
$\log_{10}(NA)$ vs. TH	1984	0			-0.1238	$p > 0.5$	22	-0.8734	$p < 0.001$	50[b]
	1986	0			-0.3254	$p > 0.1$	21	-0.8546	$p < 0.001$	51[b]
$\log_{10}(BA)$ vs. TH	1984	0			0.6673	$p < 0.001$	22	0.3079	$p < 0.05$	50[b]
	1986	0			0.5224	$p < 0.02$	21	0.3044	$p < 0.05$	51[b]

[a]All connected phytoplankton have a trophic height of 1, hence correlation must be 0. For fish, no correlations were calculated because there are too few data points (3 species in 1984, 1 in 1986). r is the correlation coefficient, p is the significance level (null hypothesis is no correlation), and n is the number of species.
[b]Isolated species excluded (6 species of phytoplankton in both 1984 and 1986, see Appendices 1A and 2A).
BM: body mass (kg), NA: numerical abundance (individuals/m³ in the epilimnion where species eat), BA: biomass abundance (kg/m³ in the epilimnion where species eat), TH: trophic height (see text).

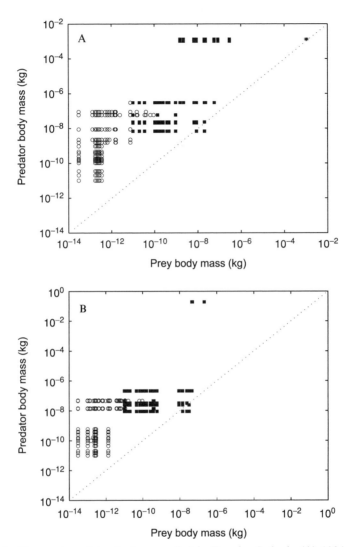

Figure 4 Prey and predator body mass (kg) in Tuesday Lake in (A) 1984 and (B) 1986, one marker for every trophic link in the unlumped food web. Cannibalistic links are excluded. Dotted line indicates equal prey and predator body mass. The links are coded according to the prey. For explanation of symbols see legend to Fig. 2. For correlations and regressions see Table 4. Figure 4A is reprinted from Cohen *et al.* (2003) with permission from the National Academy of Sciences.

Table 5 Correlations and linear least square regressions for prey and predator body size or abundance in Tuesday Lake[a]

Variables	Year	r	p	n	a	b
$\{X = \log_{10}$ (Prey BM)	1984	0.7859	p < 0.001	263	1.5598	0.8445
$\{Y = \log_{10}$ (Predator BM)	1986	0.6094	p < 0.001	233	−1.4108	0.5928
$\{X = \log_{10}$ (Prey BM)	1984	0.8832	p < 0.001	49	0.5073	0.7350
$\{Y = \text{mean}(\log_{10}$ (Predator BM))	1986	0.6885	p < 0.001	49	−0.4782	0.6440
$\{X = \log_{10}$ (Predator BM)	1984	0.9104	p < 0.001	25	−5.4342	0.7265
$\{Y = \text{mean}(\log_{10}$ (Prey BM))	1986	0.8479	p < 0.001	22	−6.5149	0.6063
$\{X = \log_{10}$ (Prey NA)	1984	0.5016	p < 0.001	263	0.8226	0.4299
$\{Y = \log_{10}$ (Predator NA)	1986	0.3434	p < 0.001	233	2.5262	0.1790
$\{X = \log_{10}$ (Prey NA)	1984	0.6489	p < 0.001	49	0.3126	0.4869
$\{Y = \text{mean}(\log_{10}$ (Predator NA))	1986	0.3805	p < 0.01	49	1.9720	0.2565
$\{X = \log_{10}$ (Predator NA)	1984	0.6899	p < 0.001	25	4.3406	0.6838
$\{Y = \text{mean}(\log_{10}$ (Prey NA))	1986	0.5458	p < 0.01	22	4.3272	0.6120
$\{X = \log_{10}$ (Predator NA)	1984	0.6807	p < 0.001	25	6.2032	0.5997
$\{Y = \log_{10}$ (Total prey NA)	1986	0.6111	p < 0.005	22	5.4433	0.6799

	Year	r	p	n	a	b
$X = \log_{10}$ (Prey NA) $Y = \log_{10}$ (Total predator NA)	1984 1986	0.6078 0.3682	$p < 0.001$ $p < 0.01$	49 49	1.5684 2.6109	0.4753 0.2786
$X = \log_{10}$ (Prey BA) $Y = \log_{10}$ (Predator BA)	1984 1986	-0.0123 0.1029	$p > 0.5$ $p > 0.1$	263 233	-4.2699 -3.7555	-0.0147 0.1574
$X = \log_{10}$ (Prey BA) $Y = $ mean($\log_{10}$ (Predator BA))	1984 1986	-0.0253 0.5000	$p > 0.5$ $p < 0.001$	49 49	-3.9616 -1.9612	-0.0174 0.3930
$X = \log_{10}$ (Predator BA) $Y = $ mean($\log_{10}$ (Prey BA))	1984 1986	0.0888 0.3987	$p > 0.5$ $p > 0.05$	25 22	-4.4113 -4.4338	0.0281 0.1525
$X = \log_{10}$ (Predator BA) $Y = \log_{10}$ (Total prey BA)	1984 1986	0.4304 0.8013	$p < 0.05$ $p < 0.001$	25 22	-2.3378 -2.1778	0.1308 0.3462
$X = \log_{10}$ (Prey BA) $Y = \log_{10}$ (Total predator BA)	1984 1986	-0.2110 0.0807	$p > 0.1$ $p > 0.5$	49 49	-3.6997 -2.7089	-0.1551 0.0293

[a] r is the correlation coefficient, p is the significance level (null hypothesis is no correlation), and n is the number of species. In the regression equation $Y = a + bX$, a is the intercept and b is the slope.
BM: body mass (kg), NA: numerical abundance (individuals/m^3 in the epilimnion where species eat), BA: biomass abundance (kg/m^3 in the epilimnion where species eat).

considered in establishing grazer-phytoplankton links (circles in Fig. 4). In only two links, a predator consumes a larger prey. Five links are cannibalistic. Trophic links are not distributed randomly in the upper triangular part of this sorted predation matrix ($p < 0.001$, along rows, columns, or diagonals, using the χ^2 approach described by Neubert et al., 2000). In the trophic-species web, for which the cascade model was originally developed, links are not distributed randomly among columns or rows ($p < 0.001$) but are among diagonals ($p > 0.5$). Because species are, with few exceptions, consumed by species larger than themselves (Fig. 4), prey body mass is positively correlated with predator body mass (Table 5). On log-log scales, prey size increases faster than predator body mass (Table 5). If the variance in log predator size decreases with increasing prey size as Fig. 4 suggests, the assumed homogeneity of variance for hypothesis testing in linear least square regression analyses would be violated. Nevertheless, the data indicate that the ratio of body masses between predator and prey decreases as prey size increases, so that predators and prey on average become more similar in size as prey size increases.

In addition, log prey size increases with mean log predator size, while log predator size increases with mean log prey size (Table 5). The slopes of both of these relationships are less than 1, as predicted by the cascade model. The larger the prey, the more similar in size its average predator; the larger the predator, the less similar in size its average prey.

The relationship between prey and predator body sizes may have implications for ecosystem dynamics. For example, the resilience (the reciprocal of return time; e.g., Harrison, 1979; Pimm, 1982) of an ecosystem could be affected (Jonsson and Ebenman, 1998) if the per capita effects between predators and their prey are correlated to the predator-prey body mass ratio (as could be expected for energetic reasons).

b. Body Size versus Trophic Height

The trophic height of a species in the Tuesday Lake food web is significantly positively related to its body size across all species (Table 4, left rear walls in Fig. 2), and negatively related to its log rank (from large to small) in body mass ($r = -0.9139$). For zooplankton (the only group for which there is sufficient variation in trophic height), the relationship between body mass and trophic height is much weaker (but still significantly positive, Table 4) than for all species.

c. Trophic Vulnerability and Generality versus Body Size

On average, trophic vulnerability decreases with increasing body size across all species ($r = -0.4305$, $p < 0.002$) while trophic generality on average increases with body size ($r = 0.4142$, $p < 0.05$). Among phytoplankton, trophic vulnerability decreases significantly with body size ($r = -0.6933$,

$p < 0.001$), but not among zooplankton ($r = -0.2224, p > 0.2$). On the other hand, trophic generality increases significantly with body size among zooplankton ($r = -0.8882, p < 0.001$).

2. Body Size and Abundance

a. Abundance-Body Size Allometry

The negative relationship between body size and numerical abundance among all species of Tuesday Lake is highly significant (Table 4, Fig. 5A, B). Most species in Tuesday Lake fall near a diagonal with slope -1 at a biomass abundance of 10^{-4} or 10^{-5} kg/m^3 in the (log numerical

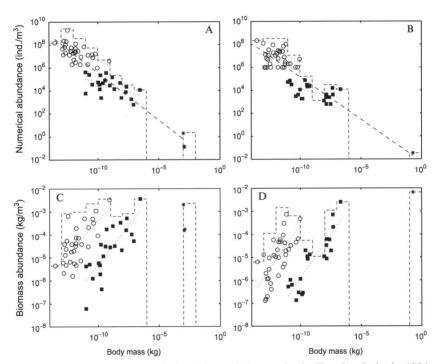

Figure 5 Body mass (kg) and abundance of the species in Tuesday Lake in 1984 (A & C) and 1986 (B & D). Numerical abundance (individuals/m^3 in the epilimnion where species eat) is shown in (A) and (B), and biomass abundance (kg/m^3 in the epilimnion where species eat) in (C) and (D). For explanation of symbols see legend to Fig. 2. Dashed line in (A) and (B) is the regression line using all species. Dotted lines are the regression lines for phytoplankton and zooplankton separately. Dash-dotted line shows the total numerical abundance (A & B) and biomass abundance (C & D; the biomass spectrum) in logarithmically increasing body mass classes. For correlations and regressions, see Tables 5 and 6.

abundance, log body mass) plane of Fig. 3. For all 56 species in 1984, including isolated species, the linear regression is

log body mass = −4.3510 − 1.0178 log numerical abundance

with 95% confidence interval (−1.14, −0.90) around the slope −1.0178. For all 57 species in 1986, including isolated species, the linear regression is

log body mass = −4.9058 − 1.0149 log numerical abundance

with 95% confidence interval (−1.1729, −0.8569) around the slope −1.0149. For both years, the confidence intervals include the slope −1 and exclude slopes equal to or greater than −3/4.

On the other hand, if the independent and dependent variables in these linear regressions are exchanged as in Fig. 5 and Table 6, then for all 56 species in 1984,

log numerical abundance = −2.6863 − 0.8271 log body mass

with 95% confidence interval (−0.92, −0.73) around the slope −0.8271. For all 57 species in 1986, the linear regression is

log numerical abundance = −2.2359 − 0.7397 log body mass

with 95% confidence interval (−0.85, −0.62) around the slope −0.7397. Only these models meet the assumptions of linear regression analysis for these data (Cohen and Carpenter, in press). For both years, the confidence intervals include the slope −3/4 and exclude the slope −1. The relationship is also significant for phytoplankton and zooplankton separately. At the 95% significance level, the slopes for these groups (Table 6) are considerably less steep than −1, −3/4, and −2/3, but not significantly different from each other.

Figure 5C and D visually confirms the quantitative conclusion (Table 6) that, across all species, the amount of biomass of each species increases only slightly from species of smaller body size to species of larger body size. Within the functional groups, however, larger species tend to have more biomass than smaller species.

The slopes of the body mass-numerical abundance relationships across all species are similar to the ones found by Marquet et al. (1990) in two rocky intertidal communities, and lie within the range reported by Cyr et al. (1997a) for 18 local aquatic communities. The slopes for these local communities lie in the range (−0.75 to −1) reported for most regional or global communities. The data of Tuesday Lake and Cyr et al. (1997a) contradict the finding (Blackburn and Gaston 1999, 1997) that at local scales, the relationship is more often polygonal and that the mean slope (−0.245) is less negative than at regional scales (mean, −0.692).

Cyr et al. (1997b) found algae and invertebrate slopes of −0.64 and −0.50, respectively (compared to −0.89 across all species) with data aggregated

Table 6 Linear least squares regression analyses of body size, abundance and trophic height in Tuesday Lake[a]

Relationship	Year	Phytoplankton a	b	n	Zooplankton a	b	n	Total a	b	n
$\log_{10}(NA) =$	1984	2.5587	−0.4071	31	1.4664	−0.3242	22	−2.6863	−0.8271	56
$a + b \times \log_{10}(BM)$	1986	4.1239	−0.2297	35	1.7306	−0.2353	21	−2.2359	−0.7397	57
$\log_{10}(BA) =$	1984	2.5586	0.5929	31	1.4663	0.6758	22	−2.6863	0.1729	56
$a + b \times \log_{10}(BM)$	1986	4.1234	0.7703	35	1.7306	0.7647	21	−2.2359	0.2603	57
$\log_{10}(BM) =$	1984				−11.7863	1.1633	22	−13.6419	1.9423	50[b]
$a + b \times TH$	1986				−11.6845	1.1453	21	−13.8240	2.0539	51[b]
$\log_{10}(NA) =$	1984				4.7254	0.1361	22	8.7400	−1.6913	50[b]
$a + b \times TH$	1986				4.5390	−0.2944	21	8.2840	−1.7060	51[b]
$\log_{10}(BA) =$	1984				−7.0609	1.0273	22	−4.9019	0.2509	50[b]
$a + b \times TH$	1986				−7.1455	0.8510	21	−5.5416	0.3479	51[b]

[a]All connected phytoplankton have a trophic height of 1. For phytoplankton no regressions were calculated for the relationships between trophic height and any other variable due to lack of variation in trophic height. For fish, no regressions were performed because there were too few data points (3 species in 1984, 1 in 1986). In the regression equation $Y = a + bX$, a is the intercept b is the slope, and n is the number of species.

[b]Isolated species excluded (6 species of phytoplankton in both 1984 and 1986).

BM: body mass (kg), NA: numerical abundance (individuals/m³ in the epilimnion where species eat), BA: biomass abundance (kg/m³ in the epilimnion where species eat) and TH: trophic height (see text).

from 18 lakes worldwide. In Tuesday Lake, the much slower decrease in numerical abundance with increasing body size within functional groups (phytoplankton and zooplankton) than across all species (Table 6) is consistent with this finding. The slopes for the plankton groups in Tuesday Lake are considerably less negative than the corresponding slopes in Cyr *et al.* (1997b) and most previously reported slopes for restricted taxonomic groups (e.g., Peters and Wassenberg, 1983; Peters, 1983). If large and rare species within functional groups (e.g., phytoplankton and zooplankton) are missing more frequently from the data than small and rare species (as may be the case when using standard microscope counting techniques; Ursula Gaedke, personal communication, 2000), this selectivity could help to explain the shallower slope within the plankton groups than across all species.

The food web of Tuesday Lake makes it possible to refine the relationship between numerical abundance and body size. In Tuesday Lake across all *consumer* species, the slope of log numerical abundance as a function of body mass was -0.67 in 1984 and -0.50 in 1986. Because these slopes are greater than -1, consumer biomass increases from the bottom to the top of the food web. The intuitive explanation is that body mass increases from the bottom to the top of the food web and for each ratio of increase in consumer body mass, there is a smaller ratio of decrease in consumer numerical abundance, so consumer biomass (which is the product of body mass times numerical abundance) increases from the bottom to the top of the food web. Why might consumer biomass increase?

In Tuesday Lake in 1984, across all species on log-log scales, the amount of resource biomass per consumer species divided by the number of consumer species utilizing each prey increases significantly ($p < 0.001$) with consumer body size. For zooplankton separately, the amount of resource biomass increases with increasing body size ($p < 0.05$). Similar results are found in 1986. The increase in available resource biomass per consumer species with increasing consumer body size could help to explain why larger consumer species have more biomass.

As predicted in section 3.4, there is: (1) a positive correlation between the slope of the body mass-numerical abundance relationship (on log-log scales) for individual pairs of consumer and resource species on the one hand, and the number of prey species of the consumer species (i.e., trophic generality) on the other ($p < 0.001$ in both 1984 and 1986); and (2) a negative correlation between the slope of the body mass-numerical abundance relationship for individual pairs of consumer and resource species on the one hand, and the number of predator species of the prey species (i.e., trophic vulnerability) on the other ($p < 0.005$ in 1984 and $p < 0.001$ in 1986). The numerical abundance of a consumer species in Tuesday Lake is less than expected if the consumer shares its prey with other consumer species, but is greater than expected if the consumer has more species of prey. This finding supports the proposition that the slope of the body mass-numerical abundance relationship within a community can be (at least partly) explained by the amount of resources available to consumers.

However, resource supply rate is not the same as the standing stock of prey biomass. Smaller organisms typically have higher energetic and biomass turnover rates than larger organisms. In line with the findings of Carbone and Gittleman (2002), and as predicted in section III.D, the slope of numerical abundance of consumers divided by an estimate of the resource productivity available to each consumer (equation 2 in Section II.A), as a function of the body mass of the consumer, is closer to $-3/4$ on log-log scales than is the relationship using the unmodified numerical abundance of consumers (Fig. 6E and F, slope -0.70 versus -0.67 in 1984 and -0.69 versus -0.50 in 1986). Each prey's estimated productivity is divided by the prey's trophic vulnerability to adjust for the number of consumers utilizing each prey species (equation 2 in Section II.A). By contrast, using the *total* productivity of all prey species in the diet of a consumer, without dividing by the number of consumers that eat each prey species, hardly changes the regression slope or the goodness of fit of the log numerical abundance regression as a function of log body mass.

The slope of numerical abundance of consumers divided by an estimate of the resource biomass available to each consumer (see Section II.A), versus the body mass of the consumer, is closer to -1 on log-log scales than the relationship using the unmodified numerical abundance of consumers (Fig. 6C and D, -0.91 versus -0.67 in 1984 and -0.86 versus -0.50 in 1986), just as Carbone and Gittleman (2002) found for carnivores from many different communities. This scaling of consumer numerical abundance also reduces the variation in abundance not accounted for by the allometric relationship (r^2 increases from 0.79 to 0.86 for the data of 1984 and from 0.70 to 0.88 for the data of 1986). As above for productivity, the biomass abundance of each prey species must be divided by the number of consumers utilizing the prey, crudely assuming equal resource use by each of the consumer species. A slope of -1 for the log-log relationship between consumer numerical abundance divided by the available prey biomass (dependent variable) and consumer body mass (independent variable) means that one unit of prey biomass supports a constant amount of predator biomass, regardless of the body size of the consumer. To prove the above, let NA_c be the consumer's numerical abundance. BM_c is the consumer's body mass, $BA_c = NA_c BM_c$ is the consumer's biomass abundance, and BA_p is the prey's biomass abundance. If we assume a slope of -1, so that

$$\log(NA_c/BA_p) = k - \log(BM_c)$$

then

$$\log(NA_c) + \log(BM_c) - \log(BA_p) = k$$

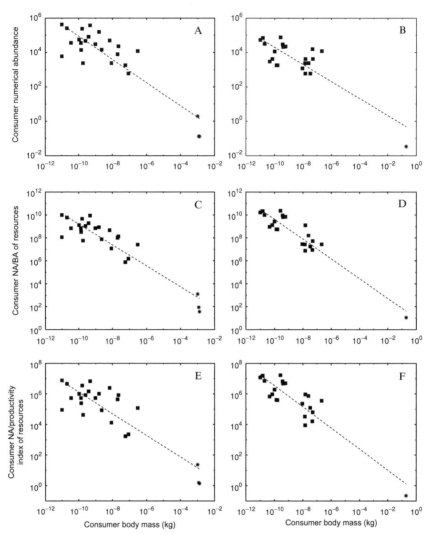

Figure 6 Body mass (kg) and abundance of the consumer species in Tuesday Lake in 1984 (A, C & E) and 1986 (B, D & F). Numerical abundance (individuals/m^3) as a function of consumer body mass (kg) is shown in (A) and (B). In (C) and (D) the numerical abundance (*NA*, individuals/m^3) of each consumer species is divided by the consumer's available resource biomass. The consumer's available resource biomass is computed as the sum, for every resource species in the diet of that consumer species, of the biomass abundance (*BA*, kg/m^3) of each resource divided by the number of consumer species that feed on that resource species (see equation (1) in Section II.A). The dimension of the ordinate in (C) and (D) is thus the number of consumers per kg of resources. In (E) and (F) the numerical abundance of each consumer species is divided by the consumer's available resource productivity.

but

$$\log(NA_c) + \log(BM_c) = \log(BA_c)$$

Therefore

$$\log(BA_c/BA_p) = k,$$

meaning that the consumer biomass per unit of prey biomass is constant, regardless of the consumer's body size. A slope greater than -1 (less steep) would mean that one unit of prey biomass will support more predator biomass, the larger the body size of the consumer. The data of Carbone and Gittleman (2002) pool many different communities while our allometric exponents are derived from observations of a single community. Both analyses indicate that the productivity of the prey determines the numerical abundance of consumers and that the conversion efficiency of prey to predator biomass is roughly similar over a wide range of predator body sizes.

Energetic mechanisms appear to explain much of the observed relation between body mass and numerical abundance in Tuesday Lake. Food web data enrich understanding of a superficially bivariate relationship between consumer numerical abundance and consumer body mass. Food webs with energy flow estimates further enrich understanding of how trophic structure affects the relationship between body mass and numerical abundance.

b. Biomass-Body Size Spectrum

The biomass-body size spectrum studied in aquatic ecology describes the amount of biomass within logarithmic size intervals, with no attention paid to the species identity of the individuals. In many aquatic and pelagic communities, the distribution of biomass is approximately uniform (e.g., Sheldon et al., 1972; 1977; Witek and Krajewska-Soltys, 1989; Gaedke, 1992). In some oceanic planktonic systems, the biomass may decrease with increasing body size (Rodriguez and Mullin, 1986). For other communities,

---

The consumer's available resource productivity is computed as the sum, for every resource species i in the diet of that consumer species, of $NA_i \times BM_i^{3/4}/n_i$, where NA_i is the numerical abundance of resource species i, BM_i is the body mass of resource species i and n_i is the number of consumer species that feed on resource species i (see equation (2) in section II.A). The dimension of the ordinate in (E) and (F) is thus the number of consumers per kg of resources per unit time. For explanation of symbols see legend to Fig. 2. Dashed lines are the regression lines using all consumer species. (A) $Y = -0.67X - 1.80$, $r = -0.8908$ (B) $Y = -0.50X - 0.67$, $r = -0.8395$, (C) $Y = -0.91X+0.10$, $r = -0.9264$, (D) $Y = -0.86X+1.00$, $r = -0.9382$, (E) $Y = -0.71X - 0.96$, $r = -0.8769$, (F) $Y = -0.69X - 0.37$, $r = -0.9030$.

the few studies available suggest a flat spectrum (Janzen and Schoener, 1968) or a spectrum with several peaks (Schwinghammer, 1981). A uniform biomass spectrum means that the amount of biomass summed over logarithmically-equal size intervals is constant over a large size range. To infer from the biomass spectrum how individual species' biomass changes with increasing body size requires additional information on how the number of species in logarithmically increasing size classes changes with body size.

Conversely, the shape of the biomass spectrum can be deduced from the relationship between numerical abundance and body size only if the frequency distribution of species by body mass is known. For example, with a log-uniform distribution of body size (i.e., constant numbers of species in logarithmically increasing size classes), the slope of the biomass spectrum is equal to 1 plus the slope of the (straight-line) relationship between log numerical abundance and log body mass. To prove the above, let log S be the log number of species as a function log body mass (log BM, here assumed to be a constant k), let log NA be the log numerical abundance as a function of log body mass log BM, and let log T be log total biomass abundance of all species as a function of their log body mass log BM. The biomass spectrum plots log T as a function of log BM. If

$$\log S = k$$

$$\log NA = a + b \log BM$$

then by definition

$$T = S \cdot NA \cdot BM$$

which implies

$$\log T = \log S + \log NA + \log BM = k + \log NA + \log BM$$
$$= (k + a) + b \log BM + \log BM = (k + a) + (b + 1)\log BM$$

Borgman (1987) reviews models of the slope of the biomass size spectrum and Vidondo et al. (1997) discuss how to analyze size spectra.

In Tuesday Lake, across all species, biomass abundance increases slightly with increasing body size (Tables 4 and 6). With a log-uniform distribution of body size, this means that the biomass spectrum should have a positive slope. Across all species, the actual spectrum (dash-dotted lines in Fig. 5C and D) has several peaks and does not seem to have a positive slope because species body masses are more log-normal or right-log skewed in distribution (Fig. 9). Within the phytoplankton and zooplankton categories, biomass of the species increases with body size (Table 6). Although the body size distributions of the phytoplankton and zooplankton are skewed as well, the slope of the biomass spectrum is positive within these groups (Fig. 5C and D). The larger species within each category tend to dominate the biomass. In summary, with no clear trend in the biomass spectrum with increasing size, the data of Tuesday Lake conform reasonably well with the flat spectrum found in other studies of pelagic communities.

c. Species Richness, Numerical Abundance and Body Size

The total number of individuals in logarithmically increasing body size classes is the numerical abundance-body size spectrum. Since biomass is the product of numerical abundance and body mass, a flat biomass spectrum implies a decreasing spectrum of numerical abundance. By dividing the log body size axis into size classes that are equal on a logarithmic scale, then counting the number of species (S_i) and the number of individuals (I_i) within each class (i), the body size distribution of species (section V.C.2) is combined with the numerical abundance-body size spectrum of individuals.

Tuesday Lake's numerical abundance-body mass spectrum (dash-dotted lines in Fig. 5A and B) decreases with increasing body size over most of the range in body size, as expected from Tuesday Lake's nearly flat biomass spectrum.

Siemann et al. (1996) studied the numerical abundance-body size spectrum and linked it to the body size distribution. In a grassland insect community, Siemann et al. (1996) found that both species richness and the number of individuals per body size class were unimodally distributed with respect to body size. The body size class with the largest number of individuals also had the largest number of species. Species richness was positively correlated to the number of individuals per body size class, roughly $S_i \propto I_i^{1/2}$. Consequently, the average number of individuals per species (A_i) scaled as $I_i^{1/2}$, and the size class with the largest number of individuals also had the largest expected numerical abundance of the species.

Tuesday Lake partially replicates the results in Siemann et al. (1996). In Tuesday Lake, species richness and numerical abundance per body size class are weakly positively correlated across all species ($r = 0.5312$, $p > 0.2$), as well as within the phytoplankton ($r = 0.7202$, $p > 0.1$) and the zooplankton ($r = 0.8141$, $p > 0.05$). Across all species, the number of species (species richness) and number of individuals peak at a similar (but not identical) body size class (Figs. 5A, B and 9). This size class is located close to the smallest body size class. To the right of this peak, with larger body sizes, the number of species and the number of individuals per size class decrease. For phytoplankton, both species richness and the number of individuals per size class are unimodally distributed with respect to body size, peaking at a similar (but not identical) intermediate body size class. Zooplankton show a similar, but less clear-cut, trend. In general, the number of species and number of individuals per size class co-vary, so that both increase up to a certain body size and then decrease with further increases in body size. These findings are qualitatively consistent with those of Siemann et al. (1996).

Contrary to Siemann et al. (1996), however, species richness is not allometrically related to the number of individuals in Tuesday Lake. The size class with the largest number of individuals and species was near the smallest size class in Tuesday Lake, not in the middle of the body size range. In the data of Siemann et al. (1996), it appeared at first that body size had no effect on the relationship between the number of individuals and the species richness per size class. However, on more careful examination of their data, the relationship between the number of individuals and species richness turns out to be a narrow parabola. Both species richness and the number of individuals per size class increase up to a certain body size and then decrease as body size increases further. In the data of Tuesday Lake, the effect of body size on this relationship is evident, since the parabola is more asymmetrical and its two legs lie further apart than in the data of Siemann et al. (1996).

The differences between these two studies may perhaps be understood by analyzing how numerical abundance and species number are associated with body size in a guild versus an entire community. Section V.C.2 suggests that log-normal or log-uniform distributions of species by body size may be a good approximation for a guild or taxonomic group. For a whole community, a right-log skewed or log-hyperbolic distribution may be more likely, and the smallest species tend to have the highest numerical abundance. When the smallest species are the most numerically abundant and the most species-rich in a community, they will dominate the numerical abundance-body size spectrum. For a guild or taxonomic group, numerical abundance and species richness could instead peak at an intermediate body size, which may increase the likelihood that species richness and total numerical abundance per body size class are allometrically related, as found by Siemann et al. (1996). Ritchie and Olff (1999) develop a theory of species diversity that predicts the distributions of body size and productivity within a group of species that use the same resource (i.e., a guild).

3. Food Web and Abundance

a. Predator-Prey Abundance Allometry
In Tuesday Lake, the numerical abundance of predators is mostly smaller than that of their prey (Fig. 7A and B). Prey numerical abundance is positively correlated to predator abundance on log-log scales (Table 5). Prey and predator numerical abundances tend to be less similar as prey becomes more abundant (Table 5). Hence, there is a larger relative difference in numerical abundance between phytoplankton and their zooplankton predators (on average) than between zooplankton and their predators.

In Fig. 7A and B, one distinct cluster of points represents phytoplankton prey (circles to the right); another represents zooplankton prey (squares to

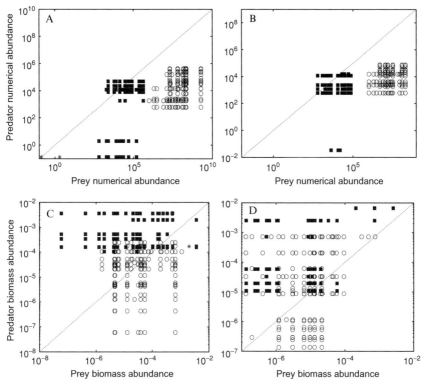

Figure 7 Prey and predator abundance in Tuesday Lake in 1984 (A & C) and 1986 (B & D). Numerical abundance (individuals/m³ in the epilimnion where species eat) is shown in (A) and (B), and biomass abundance (kg/m³ in the epilimnion where species eat) in (C) and (D). Dotted line indicates equal prey and predator abundance. Markers are coded according to the prey. Cannibalistic links have not been plotted, but other cyclic links are included. For explanation of symbols see legend to Fig. 2. For correlations and regressions see Table 4. Figure 7A and C reprinted from Cohen *et al.* (2003) with permission from the National Academy of Sciences.

the left). Within the zooplankton prey category, one cluster of data points displays a large difference in numerical abundance between prey and predator (bottom of the graph), while another displays more similar numerical abundance of prey and predator (middle of the graph). These two groups correspond, respectively, to zooplankton consumed by fish and to zooplankton consumed by other zooplankton.

No convincing relationship between predator and prey biomass abundance (Fig. 7C and D) emerges across all species (but see below), even though there is a statistically significant negative correlation between prey and predator biomass abundance (Table 5). The biomass abundance of predators that consume zooplankton was, in most cases, larger than that of their prey (squares in Fig. 7C and D). No clear trend for the biomass abundance of predators that consume phytoplankton (circles in Fig. 7C and D) could be seen.

b. Abundance versus Trophic Height

There is a significant negative correlation between trophic height and numerical abundance (projection on right rear wall in Fig. 2A and B, Table 4) across all species. For the zooplankton, the relationship is nonsignificant. The relationship between trophic height and biomass abundance (Fig. 2C and D, Table 4) across all species and for zooplankton is significantly positive.

c. Trophic Vulnerability and Generality versus Abundance

Trophic generality is weakly negatively correlated ($r = -0.1962$, $p > 0.2$) and trophic vulnerability is positively correlated ($r = 0.4210$, $p < 0.005$) to numerical abundance. With biomass abundance, trophic generality is positively correlated ($r = 0.5609$, $p < 0.005$) and trophic vulnerability is weakly negatively correlated ($r = -0.1228$, $p > 0.2$).

d. Trophic Pyramids

As described later, the three capital letters P, Z, and F in the upper right corner of Fig. 12B plot the aggregated biomass abundance of phytoplankton, zooplankton, and fish, respectively, in 1986 as a function of their values in 1984. The aggregate biomass abundance increased from fish to zooplankton to phytoplankton in 1984, but decreased along this sequence in 1986.

A descriptive pyramid may be constructed by putting species into discrete trophic levels [1 2), [2 3), [3 4), [4 5), [5 6), where, for example, the range of trophic heights [2 3) includes any species with trophic height greater than or equal to 2, up to 2.999999 (i.e., less than 3). Once species are categorized in this way by trophic height, the width of a bar can represent the sum of any characteristic that can be summed over species, such as numerical abundance or biomass abundance. Such a bar plot is merely a histogram turned on its side.

Numerical abundance decreases with increasing trophic height in Tuesday Lake in both years, but biomass abundance is much less regular as a function of trophic height (Fig. 8). In 1984, trophic level 4 has the largest total biomass abundance of all trophic levels (Fig. 8C). In 1986, the distribution of biomass abundance is hourglass-shaped with a biomass minimum on trophic level 3 (Fig. 8D). In retrospect, Figs. 1, 2, 3, and 7 all indicate a pyramid of numbers but not a (monotonic decreasing) pyramid of biomass in Tuesday Lake. Numerical abundance decreases, and biomass abundance sometimes increases, with increasing trophic height (Figs. 1 and 2, Table 4).

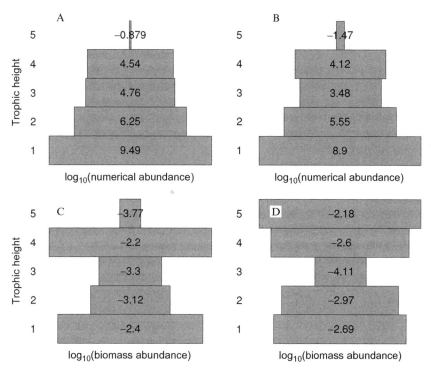

Figure 8 Total abundance by trophic height in Tuesday Lake in 1984 (A & C) and 1986 (B & D). Numerical abundance (individuals/m^3 in the epilimnion where species eat) is shown in (A) and (B), and biomass abundance (kg/m^3 in the epilimnion where species eat) in (C) and (D). Total abundance was calculated as the sum of the species abundance after species were put into discrete trophic height categories [1 2), [2 3), . . . , [4 5). The width of the bars and numbers in bars is the log$_{10}$(abundance).

We hypothesized in Section III.E that the changes in numerical and biomass abundance from one trophic level to the next could be inferred from the corresponding changes in average body mass. In 1984, the difference in average trophic level body mass is large between trophic levels 1 and 2 and between trophic levels 3 and 4, but very small between trophic levels 2 and 3 and between trophic levels 4 and 5. Our hypothesis suggests a large decrease in biomass abundance from trophic level 2 to trophic level 3 and from level 4 to level 5 as well as a small decrease, or even increase, in biomass abundance from trophic level 1 to 2 and from trophic level 3 to 4. In 1986, the difference in average trophic level body mass is large between trophic levels 1 and 2 and between trophic levels 4 and 5, but very small between trophic levels 2 and 3 and between trophic levels 3 and 4, suggesting a large decrease in biomass abundance from trophic level 2 to trophic level 3 and from trophic level 3 to trophic level 4 as well as a small decrease, or even increase, in biomass abundance from trophic level 1 to 2 and from trophic level 4 to 5. As can be seen in Fig. 8, these predictions are only partly true. Trophic levels are, however, crude descriptions of the interactions in a community. In Tuesday Lake, many species feed on multiple trophic levels. Hence, we now focus on individual consumer species.

Only 22 of 264 trophic links that are not cannibalistic connect a predator species and a prey species where predator numerical abundance exceeds prey numerical abundance (Fig. 7A and B), but 163 of 264 noncannibalistic links connect a predator species and a prey species where predator biomass exceeds prey biomass (Fig. 7C and D). Only 1 of 25 predator species has larger biomass abundance than the *total* biomass of all of its prey species.

As predicted in Section III.E, on log-log scales: (1) the ratio between predator and prey numerical abundance is positively correlated to the prey-predator body mass ratio in Tuesday Lake ($p < 0.001$ in both 1984 and 1986); and (2) the ratio between predator and prey biomass abundance is positively correlated to the predator-prey body mass ratio ($p < 0.05$ in 1984 and $p < 0.001$ in 1986). At the level of the individual consumer species, the change along a single trophic link in numerical and biomass abundance can be inferred on average from the body mass ratio of a consumer to its prey. Furthermore, the ratio of consumer biomass to an index of the available resource biomass previously described changes only slightly with increasing consumer size on log-log scales.

Therefore, the larger the relative differences in body size between a predator and its prey, the greater the ratio of consumer to resource biomass abundance will be. The amount of consumer biomass per unit of available prey biomass changes little with increasing consumer size. Despite the lack of complete success in explaining the distribution of biomass abundance across trophic levels in Tuesday Lake by average differences in body mass across trophic levels (Fig. 8), at the level of the individual consumer species the ratio of consumer to resource biomass abundance can be related to the ratio of consumer to resource body mass.

De Ruiter *et al.* (1995) and Neutel *et al.* (2002) analyzed the local asymptotic stability of Lotka-Volterra–type models of real food webs and pointed out some dynamic implications of the shape of biomass pyramids in communities. Instead of drawing the interaction strengths of such models at random from the same distribution for all species as in the work of May (1972), they estimated the interaction strengths from observed abundance data and assumptions of equilibrium feeding rates (de Ruiter *et al.*, 1994). DeRuiter *et al.* (1995) showed that this approach gives rise to interaction strengths that promote local asymptotic stability in some soil ecosystems. Neutel *et al.* (2002) showed that the increase in local asymptotic stability is caused by long trophic loops that contain relatively many weak links and can be inferred from the existence of biomass pyramids in communities. They conclude that a marked decrease in biomass with increasing trophic levels, together with predators feeding on several types of prey, caused weak links to aggregate in long loops, thereby preventing complex food webs from being unstable. That is, compared to communities where interaction strengths are distributed randomly, communities that display pyramids of biomass, and thus a characteristic pattern in the distribution of interaction strengths, are much more likely to have a low maximum loop weight, a characteristic that is shown to increase the probability of local asymptotic stability.

Raffaelli (2002, in his commentary on Neutel *et al.*, 2002) suggested that the slope of the side of the biomass pyramid in a community (assuming such a slope exists) could be an indicator of the stability of that community. Tall, thin pyramids with a high ratio of consumer to resource biomass should be less likely, he suggested, to be stable than short, relatively flat pyramids with a smaller ratio of consumer to resource biomass.

Tuesday Lake does not have a traditional pyramid of biomass when biomass abundance is summed within discrete trophic levels. The significance, if any, of the shape of the biomass distribution by trophic levels for the stability of Tuesday Lake's populations remains unclear. A remaining challenge is to try to parameterize and analyze a dynamical model of the pelagic community of Tuesday Lake.

C. Univariate Distributions

1. The Food Web

Table 3 summarizes and compares the Tuesday Lake pelagic food web in 1984 and 1986. The food web graphs of Tuesday Lake (Fig. 1A and B) show a high density of links and many species of autotrophs, fewer primary consumers, and many fewer secondary consumers.

In the unlumped web, a connected species interacts trophically with roughly 5 to 15 other species (resources and consumers). The zooplankton species are more highly connected than both the phytoplankton and the fish (since the zooplankton have both prey and predators, but the phytoplankton only predators and the fish mainly prey). In both the unlumped web and the lumped or trophic-species web, the number of resources per consumer species is greater than the number of consumers per resource species, as would be expected in a food web of pyramidal structure. The number of species, the number of food chains, food chain length, and linkage density are greater, but connectance is lower, in the unlumped webs than in the trophic-species webs.

The connectance of the food web of Tuesday Lake is considerably greater than that in the catalogues of Cohen *et al.* (1990), Schoenly *et al.* (1991) and Havens (1992), particularly when compared to webs with similar numbers of species. The connectance and number of trophic species of the Tuesday Lake food web are similar to those in the webs described by Warren (1989) and Polis (1991), but connectance is considerably higher than in the more species-rich webs of Hall and Raffaelli (1991), Martinez (1991) and Goldwasser and Roughgarden (1993). Warren (1994) reviews mechanisms affecting the connectance of food webs and the relationship between connectance and web size. In Tuesday Lake, food chains are longer (but not dramatically so) than in many previous studies.

Two empirical findings suggest that real food webs should have a pyramidal trophic structure. First, most species in communities have small bodies. Second, trophic position increases on average with body size. Consequently, food webs should have many species at low trophic heights and few species at high. This conclusion ignores the existence of parasites for which trophic position increases with decreasing body size.

The food web of Tuesday Lake differs from cascade model webs in two important aspects: the web has a pyramidal trophic structure and links are not distributed randomly among species categories, trophic levels, or among species. The pyramidal structure of the webs causes food chains to be shorter than expected under the cascade model for such a highly connected food web. For example, most intermediate species in the Tuesday Lake food web have a trophic height of 2. A food web of the same size and connectance, but in which the intermediate species have more widely varying trophic heights, should have longer food chains.

A pyramidal structure also partly explains the difference between the observed web and the cascade model webs in the distribution of links among basal, intermediate, and top species. By definition, with equal predation probabilities between any potential predator and any potential prey, a high number of basal and a low number of top species, as in the observed web, will lead to a high number of basal-intermediate links and a low number of intermediate-top links (contrary to cascade model predictions). Second, a χ^2-test of the distribution of links shows that the links are not distributed randomly ($p < 0.001$) among the species categories (basal, intermediate, and top species) in the unlumped web or in the trophic-species web. There are more observed than predicted basal-intermediate links and fewer observed than predicted intermediate-intermediate and basal-top links. If species are put into discrete trophic levels as before [1 2), [2 3), ..., [5 6), then the null hypothesis that the fraction of realized links is constant among trophic levels can be rejected by a χ^2-analysis for the unlumped web ($p < 0.001$). Compared to a random distribution, there is an excess of links between nearby trophic levels. There is also a deficit of links within the second trophic level and between the first trophic level and any level above the second. Links between distant trophic levels not involving the first level occurred approximately as often as expected. Some modifications of the cascade model (Cohen, 1990) have suggested that links could be more confined to species at nearby trophic levels, as seems to be the case here.

Schoener (1989) discussed how average vulnerability and generalization vary with the number of species in a community. He argued that the number of predator species a prey species can effectively defend against could be constrained, as well as the number of prey species a predator species can consume. Schoener (1989) predicted that vulnerability would increase with the number of species in a web, but that generalization would be unaffected by the number of species.

The same arguments predict that *within* a web, vulnerability should increase with *de*creasing trophic height, since the lower the trophic height of species the more potential consumers there are. Generality should increase and then possibly level off with increasing trophic height, since the number of potential prey species of a consumer should increase with the trophic height of the consumer, but the consumer's capacity to feed on the potential prey species is limited. The cascade model also predicts that vulnerability should decrease with increasing trophic height within a web, but does not predict any upper limit to the number of prey species a predator species can consume.

The predictions of the cascade model for the relationships between trophic height on the one hand and trophic generality and trophic vulnerability on the other agree with the observations in the food web of Tuesday Lake. Trophic vulnerability is weakly negatively correlated and trophic generality is weakly positively correlated to trophic height (but not significantly so, $p > 0.05$). Species low in the food web tend to have more predators and fewer prey than species high in the web. Across all species, generality was more weakly correlated to trophic height than vulnerability; generality does not seem to increase without limits with increasing trophic height. It is not clear from these data whether or not there is an upper limit to trophic generality. The data of Tuesday Lake do not refute the hypothesis of unconstrained vulnerability and constrained generality within a food web, but are hardly decisive.

a. The Distribution of Species

The number of basal species slightly exceeds the number of intermediate species, which is far greater than the number of top species in the unlumped web of Tuesday Lake (Table 3). Basal, intermediate, and top species correspond, with a few exceptions, to phytoplankton, zooplankton, and fish. However, in the trophic-species (lumped) web, the number of intermediate trophic species exceeds the number of basal trophic species, which is greater than the number of top trophic species.

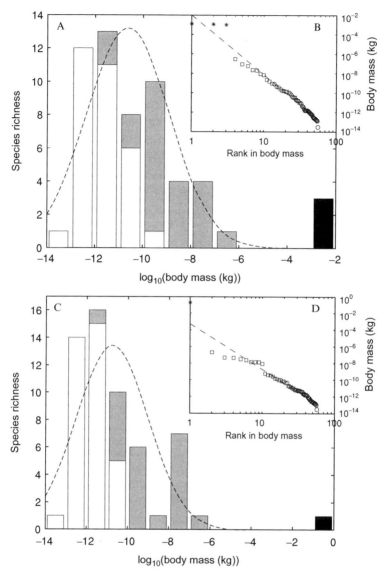

Figure 9 (A & C) The frequency distribution of the number of species (species richness) by $\log_{10}$ body mass (kg) in Tuesday Lake in (A) 1984 and (C) 1986. White bars = phytoplankton, gray bars = zooplankton, black bars = fish. Dashed line is the log-normal distribution with the same mean and variance as the observed distribution with fishes excluded. (B & D) Body mass vs. the rank in body mass for the species in Tuesday Lake in (B) 1984 and (D) 1986. Rank goes from largest body mass to smallest. Dashed lines are the ordinary least squares regression lines, using all species. (B) $Y = -6.16X - 1.99$, $r = -0.9861$, (D) $Y = -5.43X - 3.27$,

The cascade model (Section III.A.1) predicts equal numbers of basal and top trophic species. The observed distribution of trophic species in Tuesday Lake (Table 3) is significantly different ($p < 0.01$, χ^2-test) from that predicted by the cascade model, primarily because of a large number of basal trophic species.

b. The Distribution of Trophic Links

In Tuesday Lake, the number of links between basal and intermediate species is approximately twice the number of links between intermediate and intermediate species. These numbers are much greater than the number of links between intermediate and top species. Links between basal and top species are absent. The trophic-species webs have more basal-intermediate links, and fewer basal-top and intermediate top-links, than predicted by the cascade model (Table 3).

c. The Distribution of Chain Lengths

Food chain lengths are more or less normally distributed in both the un-lumped web and in the trophic-species web. However, observed food chains are on average (Table 3) much shorter than expected ($p < 0.001$, using the normal deviate as described in Zar, 1999) if links were distributed randomly among the species. To make this comparison, observed noncannibalistic links were randomly redistributed in the upper triangular part of the predation matrix and the observed mean chain length was compared to the distribution of simulated means in 100 replicates.

2. Body Size

The distribution of log body mass of the species in Tuesday Lake (Fig. 9A and C) is skewed and deviates significantly from a normal distribution ($p < 0.001$). The absence of species between the largest zooplankton species and the fish gives a wide gap in the distribution. Other potential gaps are located between 10^{-9} and 10^{-8} kg in both years and between 10^{-10} and 10^{-11} kg in 1984. Holling (1992) reviewed mechanisms that may lead to gaps in the body size distribution in communities. If fish are excluded, the null hypothesis of normality of log body size cannot be rejected ($p > 0.1$). The distributions of log body size do not deviate significantly from normal distributions

---

$r = -0.9714$. Y is $\log_{10}$(body mass) and X is $\log_{10}$(rank in body mass). A straight-line relationship between Y and X represents a power-law distribution of body mass. For explanation of symbols, see legend to Fig. 2.

for phytoplankton and zooplankton separately ($p > 0.1$ and $p > 0.25$, respectively).

The slope of the right tail of the body size distribution in Tuesday Lake (fish excluded) is approximately linear on log-log scales and considerably less steep than $-2/3$ both for phytoplankton and zooplankton combined and separately. While the shape and slope of the tail of any distribution can be affected by the choice of histogram intervals, and the tail of the body size distribution need not necessarily be linear on log-log scales (Loder et al., 1997), these possibilities are not of concern here. Tuesday Lake has relatively fewer small species and/or more large species than, for example, the studies reviewed by May (1986). We predict that combining data from a number of similar pelagic systems would add relatively more small species and would decrease the slope of (i.e., make steeper) the right tail of the distribution.

Another way of looking at the distribution of body sizes is to plot the body mass data by their rank. In Tuesday Lake, both allometric and exponential models are good approximations to the relationship between body mass and body mass rank. However, an allometric or power-law model (Fig. 9B and D) fits slightly better than an exponential model for all species ($r = -0.9861$ versus $r = -0.9064$) and for phytoplankton only ($r = -0.9805$ versus $r = -0.9543$), but not for zooplankton only ($r = -0.9620$ versus $r = -0.9855$).

The nearly linear relationship between log body mass and log rank in body mass argues against the log-normality of the distribution of body sizes. If body mass is allometrically related to rank ($BM \propto \text{rank}^{\alpha}$), then the frequency distribution of body mass may be more log-hyperbolic than log-normal. This pattern in Tuesday Lake differs from that reported for North American land mammals (Brown and Nicoletto, 1991), which suggests nearly log-uniform body size distributions at local geographical scales.

3. Abundance

Figure 10A and B shows the distributions of numerical abundance and the rank-numerical abundance relation in Tuesday Lake. The numerical abundance of phytoplankton exceeds that of zooplankton on average by approximately 3 orders of magnitude and their distributions do not overlap (Appendices 1A and 2A). The excess of zooplankton over fish numerical abundance is even greater. Across all species, the distribution of numerical abundance consists of three separate distributions, and the deviation from a normal distribution of log numerical abundance is statistically significant ($p < 0.001$). The distributions of phytoplankton and zooplankton abundance do not deviate significantly from log-normal distributions, separately ($p > 0.5$)

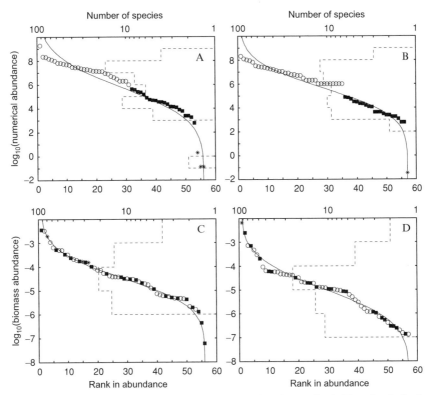

Figure 10   Abundance vs. the rank in abundance for the species in Tuesday Lake in 1984 (A & C) and 1986 (B & D). Rank goes from greatest abundance to smallest. Numerical abundance (individuals/m^3 in the epilimnion where species eat) is shown in (A) and (B), and biomass abundance (kg/m^3 in the epilimnion where species eat) in (C) and (D). For explanation of symbols see legend to Fig. 2. Solid line is the expected rank-abundance relationship assuming that abundance is lognormally distributed. (Drawing 10,000 values from a normal distribution with the same mean and variance as the observed distribution of log abundance and plotting abundance vs. rank in abundance produced the line.) Dashed line is the (log scale) frequency distribution of the number of species (top horizontal axis) by (log scale) numerical abundance in (A) and (B), and by (log scale) biomass abundance in (C) and (D).

or combined ($p > 0.2$). A log-normal distribution of numerical abundance across all species is a reasonable, but not perfect, approximation of the data (solid line in Fig. 10A, using the same mean and variance of log numerical abundance as that in the observed distribution). The observed decrease in log numerical abundance with increasing rank is close to linear within the two plankton categories, indicating departures from log-normality but, on the contrary, agreement with a power-law distribution. The slope of rank-log numerical abundance is shallower for phytoplankton than for zooplankton. Abundance declines more slowly with increasing rank in phytoplankton than in zooplankton. At least in 1984, the plot of log numerical abundance by rank is nearly concave, as predicted in Section III.C for pyramidal food webs. Thus, as predicted, the pyramidal shape of the food web of Tuesday Lake is reflected in the rank-numerical abundance relationship.

Figure 10C and D shows the distributions of biomass abundance and the rank-biomass abundance relations in Tuesday Lake in 1984 and 1986. The ranks of phytoplankton, zooplankton and fish by biomass abundance are mixed so that there are no gaps in the distribution. The difference in biomass abundance between phytoplankton and zooplankton is not significant (one-way ANOVA: $p = 0.33$). A few species dominate the distribution of biomass abundance. The frequency distribution of biomass abundance does not differ significantly from a log-normal distribution, considering all species together (dashed line in Fig. 10C and D, $p > 0.7$) or phytoplankton and zooplankton species separately ($p > 0.25$ and $p > 0.69$, respectively). The observed rank-biomass abundance relationship conforms well, by visual inspection, to that predicted using a log-normal distribution of biomass with the same mean and variance of log biomass abundance as that in the observed distribution (solid line in Fig. 10C and D).

Other models of the rank-numerical abundance relationship include the log-series and broken-stick distributions (for review, see May, 1975). A log-normal distribution is symmetrical and sigmoid in shape when log abundance is plotted as a function of rank, whereas a log-series distribution displays a linear decrease in log abundance with increasing rank. The broken-stick distribution, intermediate between these two, shows an almost linear decrease in log abundance over a large part of the range in rank.

Across all species in Tuesday Lake, the rank-log abundance relationship is not linear for numerical or biomass abundance. However, the relationship is very close to linear for numerical abundance within the species categories (phytoplankton and zooplankton), suggesting a broken-stick or log-series relationship within these species categories.

These results are in line with expectations. A broken-stick relationship is expected when a homogenous group of species (e.g., a guild) divides a limiting resource (or niche space) randomly and each species' numerical abundance is proportional to its share of the resource (MacArthur, 1957), although the same relationship can be derived from quite different assumptions (Cohen, 1968). Empirically, the broken-stick distribution of numerical abundance is usually found in small, homogeneous taxa of similar body size where the numerical abundance is thought to be governed by one (or a few) factors or limiting resources. The log-series distribution is expected when abundance is governed by one or few factors (as in the broken-stick model), but where the partitioning of the resource or niche space is highly hierarchical. Harsh environments with low resource levels or high levels of disturbance are thought to lead to log-series distributions. This distribution is also expected in early successional stages of communities (Gray, 1987). The log-normal distribution applies to larger, more heterogeneous groups where many independent factors affect the abundance. Log-normal distributions of numerical abundance should be associated with undisturbed whole communities in equilibria, where competitive species interactions are abundant (May, 1975; Tokeshi, 1993; but see Nummelin, 1998; Watt, 1998; and Section VI.H).

In Tuesday Lake, the numerical abundance of phytoplankton and zooplankton is not log-normally but rather broken-stick distributed or concave as a function of rank in abundance, and the biomass abundance of all species is approximately log-normally distributed. This result suggests that the numerical abundance of phytoplankton and zooplankton may be affected by a few factors only, but the numerical and biomass abundance across all species may be determined by many independent factors.

The four trophic groups of Tuesday Lake—namely, phytoplankton, herbivorous zooplankton, carnivorous zooplankton, and fish—are fairly distinct with respect to body size. Since numerical abundance is well correlated with body size, these groups separate in a rank-numerical abundance plot. In other communities as well, the rank-numerical abundance plot could be composed of several broken-stick-like distributions. As the number of trophic groups increases and their body sizes overlap, the relationship may increasingly resemble a log-normal distribution.

The relationship between body mass and body mass rank (Section V.C.2) can be used to predict the numerical abundance of the species. Accepting an allometric relationship $BM_i = \alpha i^\beta$ between body mass BM_i and body mass rank i (Fig. 9B and D) and accepting that numerical abundance is allometrically related to body mass (Fig. 5A and B), it follows that numerical abundance is allometrically related to body mass rank as well. The data of Tuesday Lake (Fig. 11A and B) indicate a reasonable approximation to this prediction ($r = 0.9026$). Further, if body mass and numerical abundance are allometrically related to body mass rank, then so is biomass abundance, with an exponent that is determined by the exponents for body mass and numerical abundance. With $BM \propto \text{rank}^{-6.16}$ and $NA \propto \text{rank}^{+5.08}$ in 1984, it follows that $BA \propto \text{rank}^{-1.08}$. That the exponent -1.08 is negative is a prediction that higher ranked (smaller bodied) species will have smaller biomass abundance. As predicted, in Tuesday Lake, smaller species (with higher body mass ranks) tend to have smaller biomass than larger species (Fig. 11C and D).

As approximate linear or allometric relations are compounded by these theoretical calculations, the scatter of data points with respect to the predicted linear relationships noticeably increases (e.g., compare successively

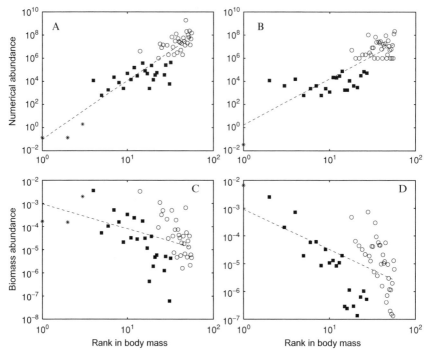

Figure 11   Numerical abundance (A and B, individuals/m^3 in the epilimnion where species eat) and biomass abundance (C and D, kg/m^3 in the epilimnion where species eat) of the species in Tuesday Lake in 1984 (A and C) and 1986 (B and D) plotted as a function of the rank in body mass (where rank goes from largest body mass to smallest). For explanation of symbols see legend to Fig. 2. Dashed lines are the ordinary least squares regression lines, using all species, with $Y = \log_{10}$(abundance) and $X = \log_{10}$(rank in body mass). (A) $Y = 5.08X - 1.02$, $r = 0.9026$ (B) $Y = 3.99X + 0.23$, $r = 0.8344$, (C) $Y = -1.08X - 3.02$, $r = -0.4331$, (D) $Y = -1.45X - 3.04$, $r = -0.5187$.

Figs. 5A and 9B with 12A and finally 12C). A theory of variability is needed along with a theory of expected relationships.

VI. EFFECTS OF A FOOD WEB MANIPULATION ON COMMUNITY CHARACTERISTICS

This section compares Tuesday Lake in 1984 and 1986, one year before and after the 1985 manipulation that removed three species of fish and introduced another species of fish (Section IV.A). Some conclusions on the effect of a food web manipulation on community patterns are suggested.

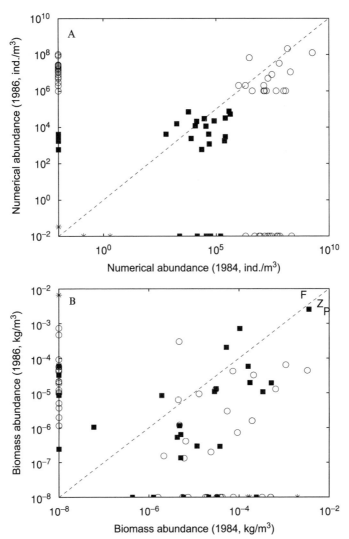

Figure 12 The abundance of the species in Tuesday Lake in 1984 and in 1986. (A) numerical abundance (individuals/m³ in the epilimnion where species eat) and (B) biomass abundance (kg/m³ in the epilimnion where species eat). For explanation of symbols see legend to Fig. 2. Symbols on the line $Y = 10^{-2}$ in (A) and $Y = 10^{-8}$ in (B) denote species that were present in 1984 only, those on the line $X = 10^{-2}$ in (A) and $X = 10^{-8}$ in (B) denote species present in 1986 only. Dashed line indicates equal abundance in 1984 and 1986. Letters in (B) indicate the total biomass abundance of all phytoplankton species (P), all zooplankton species (Z) and all fish species (F).

A. Species Composition and Species Turnover

The total number of species remained almost the same, but the species composition changed (Appendices 1A and 2A, Table 3). There were more species of phytoplankton in 1986 than in 1984, but fewer zooplankton and fewer fish species. Most species were phytoplankton in both years. Approximately 60% of the species present in 1984 were also present in 1986. Of the number of phytoplankton species present in a particular year, approximately one-half occurred in that year only. Approximately one-quarter of the zooplankton species were unique to a particular year. Thus, the zooplankton species present were more similar between 1984 and 1986 than the phytoplankton species present. The turnover rate of the phytoplankton species was 32 of 49, or approximately 65.3% (14 species disappeared and 18 species appeared between 1984 and 1986). The turnover rate of zooplankton species was 11 of 27, or approximately 40.7% (6 species disappeared and 5 species appeared).

We lack a theory for what appear to be high rates of species turnover. We do not know if the apparent changes in species composition result from insufficient sampling in both years, or from real local extinctions and introductions, or from both.

B. Food Web, Body Size, and Abundance

Unlike 1984, body mass and trophic height explain approximately equal amounts of variation in numerical abundance in 1986 (Table 4). Multiple regression slightly increases the proportion of variation explained. The food web manipulation decreased the predictive power of body mass for abundance to about the same level as that of trophic height.

C. Food Web and Body Size

Most patterns that involve body size and trophic interactions were similar in 1984 and 1986. In both years, predators generally consumed smaller prey (Fig. 4), trophic vulnerability decreased, and trophic height increased on average with body size (Fig. 2). Unlike 1984, trophic generality did not increase significantly ($p > 0.1$) with body size in 1986. Geometric mean predator size increased more slowly (but not significantly so, $p = 0.2057$) with prey size in 1986 than in 1984 (Table 5), and geometric mean prey size increased with predator size more slowly in 1986 than in 1984, but again not significantly more slowly ($p = 0.1368$). As predicted in Section III.A.2, trophic height is better predicted by log rank in body mass in 1984 than in 1986 ($r_{84}^2 = 0.84$ versus $r_{86}^2 = 0.76$).

D. Food Web and Abundance

In both years, the numerical abundance of predators was generally smaller than that of their prey. But in 1986, fewer zooplankton species than in 1984 were consumed by a predator with a much lower numerical abundance (squares in lower middle part of Fig. 7A and B). The fish predator in 1986 consumed fewer zooplankton species than the fish predators in 1984. In both years, the biomasses of predators consuming zooplankton were mostly higher than that of their prey (squares in Fig. 7C and D). In 1986, there was more variation in both predator and prey biomass (i.e., the squares are more scattered in Fig. 7D than in C). The same is true for the interaction between phytoplankton and their consumers (circles in Fig. 7C and D).

In both years, the food web of Tuesday Lake showed an approximate pyramid of numbers for species in discrete trophic levels (Fig. 8A and B) and for phytoplankton, zooplankton, and fish. Unlike 1984, fish biomass in 1986 exceeded zooplankton biomass, which exceeded phytoplankton biomass (letters F, Z, and P in Fig. 12). For species in discrete trophic levels, the bar plot is hourglass-shaped in 1986 (Fig. 8D). In 1986, of the 236 trophic links that are not cannibalistic, only 17 connected a predator and a prey where predator numerical abundance exceeded prey numerical abundance (3 of these links are involved in cycles), but 157 of 236 noncannibalistic links connected a predator and a prey where predator biomass abundance exceeded prey biomass abundance (3 of these links are involved in cycles). However, only three of 22 predators have larger biomass abundance than the *total* biomass of their respective prey. Thus, as in 1984, numerical abundance of consumers was in general less than that of their prey, whereas biomass abundance of consumers can be either larger or smaller than the biomass abundance of their resources. In general, the biomass abundance of a consumer was smaller than the total biomass of its prey.

E. Body Size and Abundance

The shapes of the abundance-body size relationships are similar in both years. The slope of log numerical abundance as a function of log body size is less steep in 1986 than in 1984 for phytoplankton, zooplankton, and all species (Table 6), but the differences in slope are not significant among the phytoplankton ($p = 0.16$) or zooplankton ($p = 0.28$) or for all species

($p = 0.12$). The abundance-body size spectra (dash-dotted lines in Fig. 5) are also similar in shape in the two years.

Marquet *et al.* (1990) analyzed the effect of human disturbance on the abundance-body size relationship in two rocky intertidal communities. Despite considerable effects of the perturbation on the species composition and food web, the abundance-body size relationship was not influenced. Similarly, in Tuesday Lake, the slopes of the numerical abundance-body size relationship in 1986 and 1984 (Table 6, Fig. 5A and B) are not significantly different.

We speculated in Section III.A.2 that body size and abundance could be better predicted by the rank in body mass in a community before than after a disturbance. For Tuesday Lake, this prediction holds on log-log scales for body mass ($r^2_{84} = 0.97$ versus $r^2_{86} = 0.94$, Fig. 9B and D) and numerical abundance ($r^2_{84} = 0.81$ versus $r^2_{86} = 0.70$, Fig. 11A and B) but not for biomass abundance ($r^2_{84} = 0.19$ versus $r^2_{86} = 0.27$, Fig. 11C and D). However, since the amount of variation in log biomass abundance that is explained by variation in log rank in body mass is very low in both years, this relationship may not be appropriate for detecting effects of disturbance.

F. Food Web

The food webs of Tuesday Lake in 1984 and 1986 are on the whole very similar. However, the unlumped food web in 1986 is a bit less connected, has somewhat fewer trophic links and many fewer food chains and slightly shorter food chains on average than the food web in 1984 (Table 3). Similarly, in the trophic-species web, connectance and the number of trophic links and food chains are lower, and food chains shorter on average, in 1986 than in 1984. Unlike 1984, the food web of 1986 has a few links between basal and top species because, in 1986, one species of zooplankton (*Chonochilus* colonial) was not consumed by any other species.

Connectance decreased between 1984 and 1986, mainly because the three zooplanktivorous fishes removed had a higher connectance to the rest of the web than the introduced, mainly piscivorous, species. The planktivorous fishes consumed 9.7 other species on average, while the piscivorous fish consumed 3 other species. For all other categories, the differences between years were minor.

Unlike 1984, in 1986 the observed distribution of *trophic* species in the categories basal, intermediate, and top (Table 3) was not significantly different ($p > 0.25$, χ^2-test) from that predicted by the cascade model. Furthermore, the null hypothesis of equal fractions of realized links among the different categories could not be rejected ($p > 0.25$) for the trophic-species web in 1986. In neither year were the trophic links randomly distributed in

the *unlumped* webs, neither among trophic levels nor among basal, intermediate, and top species. For both the unlumped web and the trophic-species
webs, the χ^2-values were considerably smaller in 1986 than in 1984, suggesting that the manipulation of the food web to some extent randomized the
distribution of trophic links. As in 1984, the trophic links are not distributed
randomly among the species in the upper triangular part of the predation
matrix ($p < 0.001$) in the unlumped web.

G. Body Size

The distributions of body size in Tuesday Lake (Fig. 9A and C) are similar
and right-log skewed in both years (but do not deviate significantly from a
log-normal distribution for phytoplankton or phytoplankton and zooplankton combined). Contrary to 1984, the distribution of log body size of
zooplankton in 1986 shows a significant deviation ($p < 0.005$) from normality. Furthermore in 1986, the larger zooplankton species are more horizon-
tally positioned in the rank-body size relationship (Fig. 9B and D) than
in 1984.

Holling (1992) proposed that clumps in body size distributions may reflect
discontinuities in habitat texture and do not result from, for example,
historical or trophic factors. Raffaelli et al. (2000) tested the proposition of
Holling (1992) by examining the sensitivity to perturbations of the body size
distribution in a benthic invertebrate community. Despite large-scale
changes in the composition of the community and the abundance of the
organisms as a result of the treatments, the locations of the gaps and clumps
in the body size distributions were little affected. This finding supports the
view that habitat architecture may be responsible for the shape of the body
size distribution. That the body size distribution of Tuesday Lake was
changed little by the manipulation could indicate that, despite a well-docu-
mented "trophic cascade," trophic factors alone are not the dominating
force affecting the distribution of body sizes in this community. This conclusion is corroborated by the analyses of Havlicek and Carpenter (2001), who
found that clumps and gaps in plankton size distributions were robust
to food web perturbations.

H. Abundance

Total biomass abundance of all species is similar in 1984 and 1986 (1.51×10^{-2}
kg/m^3 versus 1.25×10^{-2} kg/m^3). The 1984 total phytoplankton biomass is
larger (7.44×10^{-3} kg/m^3 versus 2.19×10^{-3} kg/m^3), the total zooplankton
biomass slightly larger (5.38×10^{-3} kg/m^3 versus 3.68×10^{-3} kg/m^3), and the

total fish biomass less (2.32×10^{-3} kg/m^3 versus 6.64×10^{-3} kg/m^3) than the corresponding biomass in 1986. An unexpected finding is that, both before and after the 1985 manipulation, the biomass abundance of these three categories of organisms were all within less than one order of magnitude of each another, despite the variation in biomass abundance of individuals species over roughly six orders of magnitude.

Most single species had lower numerical and biomass abundance in 1986 than in 1984 (Fig. 12). For phytoplankton, both log numerical abundance and log biomass abundance of the species are greater in 1984 than in 1986 (one-way ANOVA: $p < 0.02$). For zooplankton, log numerical abundance of the species, but not log biomass abundance, is greater in 1984 than in 1986 (one-way ANOVA: $p < 0.02$). Species that were present in both 1984 and 1986 are, on average, less abundant in 1986 than in 1984, in both numerical abundance (binomial test, $p = 0.002611$) and biomass abundance (binomial test, $p < 0.001$). However, two species of phytoplankton (*Chromulina* sp. and *Dinobryon cylindricum*) and four species of zooplankton (*Ascomorpha eucadis, Daphnia pulex, Holopedium gibberum,* and *Keratella testudo*) had a higher numerical and biomass abundance in 1986 than in 1984. One species of phytoplankton (*Synedra* sp.) had a higher numerical abundance, but not biomass abundance, in 1986 than in 1984.

Relative abundance changed also, but by less than one order of magnitude, for the aggregated major groups. The ratio of phytoplankton biomass to zooplankton biomass to fish biomass (P:Z:F) is 1:0.72:0.31 in 1984 and 1:1.68:3.04 in 1986. The pelagic community is dominated by phytoplankton in 1984 and by fish in 1986. The ratio of isolated species biomass (all phytoplankton) to total phytoplankton biomass is 0.46 in 1984 and 0.071 in 1986. The biomass ratio of small zooplankton (<0.001 mm) to large zooplankton is 0.43 in 1984 and 0.12 in 1986. Larger bodied zooplankton decreased much less than smaller bodied zooplankton. The biomass of isolated phytoplankton decreased more between 1984 and 1986 than the biomass of nonisolated phytoplankton. In summary, between 1984 and 1986, a community with a large amount of "inedible" phytoplankton and a significant amount of small zooplankton shifted to a community dominated by one species of fish, with a much smaller amount of both inedible phytoplankton and small zooplankton. Decreasing absolute abundance for both phytoplankton and zooplankton could result from some abiotic factor. Changes in relative abundance are more likely due to biotic factors.

The slope and shape of the rank-abundance curves (Fig. 10) are similar in 1984 and 1986. The 1984 curve consistently lies above the 1986 curve for biomass abundance, and slightly above for numerical abundance for most ranks, because the abundance in 1984 was greater than in 1986 for most species. The slope of the rank-biomass relationship is initially steeper in 1986 than 1984, mainly due to the dominance of the introduced bass.

The difference in slope between the years is gradually reduced as more species are included.

In communities that have been disturbed, a few dominating, disturbance-resistant, or resilient species may account for the majority of the individuals present. Some studies have proposed that more even or log-normal distributions are typical of undisturbed, species-rich systems (May, 1975; Gray, 1987; Tokeshi, 1993) and that a lack of fit of species' numerical abundance to a log-normal distribution could indicate ecosystem disturbance (Hill *et al.*, 1995). However, Nummelin (1998), analyzing data of forest floor vegetation and four insect groups in logged and unlogged rainforest sites in Uganda, found no support for the hypothesis that undisturbed, but not disturbed communities, are characterized by log-normal distributions of abundance. Irrespective of whether they came from disturbed or undisturbed sites, the distributions of numerical abundance fitted a log-normal distribution. Watt (1998) criticized the use of species-abundance models as indicators of ecosystem disturbance by pointing out that conclusive empirical support for the hypothesis is lacking, that there may be far better ways to assess whether a community has been disturbed or not and that the method provides no quantitative measure of the degree of the disturbance.

In Tuesday Lake, the relative abundance of species was less even in 1984 before the manipulation than in 1986. Regardless of the effects of disturbance on the distribution of abundance, *which* species appear or disappear and *which* species increase or decrease significantly, as a result of a disturbance, could be more interesting and informative than a change (or lack of change) in the shape of a rank-abundance relationship, which ignores species identity or characteristics other than abundance.

I. Conclusions Regarding the Manipulation

The manipulation of the fish species in 1985 gives this study a comparative aspect with the advantage that major parts of the system remained the same before and after the intervention. If we were to compare Tuesday Lake with say, a forest, there would be no way of knowing which of the many differences between the two systems were responsible for any difference in community characteristics observed.

The manipulation produced at most minor differences between 1984 and 1986 in the relationships analyzed in Sections V and VI. The food webs, the rank-abundance relationships, and abundance-body size relationships are similar in shape. In both years, the distribution of body size was right-log skewed, the biomass spectrum across all species flat, and consumers were with few exceptions larger and less numerically abundant than their prey. Species composition changed, as did the numerical and biomass abundance of the species and relative abundance among species categories.

At least three possibilities could explain why the manipulation did not have major effects on community characteristics. First, the lake could be constantly perturbed by natural climatic and biotic fluctuations, so that the human manipulation in 1985 was not different in kind from major upheavals experienced prior to 1984. According to this possibility, nothing much changed because the manipulation was business as usual. Second, the manipulation was not severe enough to affect the community characteristics analyzed here. Third, the major effects of the disturbance take longer than one year to appear.

Testing the first and third alternatives requires a longer time series of detailed observations than the two years available here. We do not know if 1986 represents a transient stage or a new steady state of the system. If the lake is constantly perturbed, all states are transient. The question of dynamics lies beyond the data, and therefore beyond the scope, of the article. Our results do not bear directly on questions of complexity and stability.

If the second alternative is correct, then even though the species changed, many of the constraints on species imposed by ecological interactions did not change. The range in body size and abundance of the species in a whole community is very large compared to the variation in these variables that a food web manipulation may cause for individual species. Effects of a perturbation that is noticeable at the species level could make little difference at the community level.

VII. DATA LIMITATIONS AND EFFECT OF VARIABILITY

The data on Tuesday Lake used here have at least seven limitations that could affect the relationships analyzed.

First, the community boundaries are defined to include the pelagic food web of Tuesday Lake, and to exclude the littoral zone and *Sphagnum* bog that surround the lake. Feedbacks between the littoral and pelagic zones are well documented in other lakes (Boers *et al.*, 1991; Persson *et al.*, 1992; Carpenter *et al.*, 1992). Since Tuesday Lake is a small lake, it has a large ratio of perimeter to lake area and a large ratio of surface to volume. Stomach contents showed that some food of fish was littoral (and even terrestrial in a few cases). These extrapelagic sources may help to explain the high biomass of fish relative to their pelagic food base, and inclusion of the lake's benthic fauna might fill some of the gap in the body size distribution (Fig. 9). The littoral zone in Tuesday Lake could be important. On the other hand, the littoral zone is small and sparsely vegetated, macrophytes are nearly absent, and the few invertebrates (e.g., dragonfly larvae and beetles) are associated with the bog edge. We infer a low benthic production and a minor role for the littoral zone when compared to many other lakes.

Second, the community is incompletely described within its defined boundaries. Two potentially important groups are missing: microbes and parasites. Molecular methods of determining bacterial diversity were not available at the time of this study. Studies of protozoa in Tuesday Lake began in 1988 (Pace, 1993). Bacteria are consumed by mixotrophic phytoplankton, protozoans, and zooplankton. Protozoans are consumed by zooplankton. These feeding relationships constitute the microbial loop, which may be important for nutrient recycling (see Stockner and Porter, 1988; Porter *et al.*, 1988 for reviews). Including the microbial food web may affect the zooplankton body size distribution and related measures, and will probably affect the trophic positions of the zooplankton. Better understanding of the microbial food web in lakes and its linkages to the metazoan food web reported here is an important topic for future research.

Parasites are potentially important regulators of the numerical abundance of species, but are absent from most food web descriptions (except Huxham *et al.*, 1995; Memmott *et al.*, 2000). Since parasites can affect many food web properties (e.g., create looping and increase chain lengths), incorporation of these organisms in food web descriptions is an important goal for future studies (Marcogliese and Cone, 1997).

Despite the extensive sampling, our data are a sample and other species than microbes and parasites may have been missed. However, sampling efforts were similar in the two years, and the same microscopist counted the phytoplankton in both years. Any differences in phytoplankton between the two years are not likely to be an artifact of sampling intensity or analytical bias.

Third, although a community changes in time, the data are static. They represent averages over some time and space. The food web is an accumulated web, not a snapshot of the pelagic community of Tuesday Lake. In reality, many species shift diets in response to their developmental stage and to changing prey availability. The abundance data represent seasonal averages during summer stratification (May to September). In reality, there is a succession of phytoplankton and zooplankton species during a year. Different species gain dominance and peak at different times. The body sizes are average sizes. In reality, a species is composed of a mix of juvenile and adult individuals. The body size of some organisms may range over several orders of magnitude during the growth of individuals. The average depends on the age structure of the population. If body mass increases monotonically with age, then a slowly growing or declining population has a higher concentration of older individuals and therefore a greater mean body mass than a rapidly growing population, which will have a higher proportion of young individuals even if the schedule of body mass as a function of age is the same in both populations. Further, the hypothesis in the previous sentence that body mass increases monotonically with age is not universally valid; on the contrary, individuals in some species shrink on starvation. Our analysis ignores all these complications of change.

An alternative to the approach taken here would be to have data averaged over smaller intervals of time and space. But weekly and even daily data are also averages. To obtain and to analyze temporally better resolved data of the type analyzed here are challenges for future studies (see Schoenly and Cohen, 1991; Closs and Lake, 1994; Tavares-Cromar and Williams, 1996 for studies of temporal variation in food web structure). The connectedness of different static descriptions of a community applies equally well to dynamic data, and dynamic data create many new possibilities for interesting patterns and relationships.

Fourth, not all the data analyzed were independently obtained. Since relative sizes of consumers and potential prey were used in some cases to infer trophic relations, any pattern in the distribution of trophic links (such as the food web) or relationships involving trophic relations and body size (such as predator-prey size relationships) must be interpreted cautiously. Relationships involving body size and abundance (such as rank-abundance or abundance-body size allometries) are unaffected by inferences about the food web.

Fifth, the area or volume where the zooplankton species of Tuesday Lake feed, namely the epilimnion, is not identical to the area or volume where the species live, which is about six times deeper. In which volume should abundance be expressed? We chose here to express the abundance of all species as the number of individuals per cubic meter of water in the epilimnion, where the trophic interactions take place. (The zooplankton migrate daily to the epilimnion to feed.) The zooplankton species live and were sampled in a water volume that is larger by a factor of 6. Appendices 1 and 2 express zooplankton concentrations in the larger volume where they live. For all statistical calculations reported here, we multiplied (only!) zooplankton abundance by 6 to convert the counts of zooplankton to numerical abundance per cubic meter of epilimnion. This adjustment will not qualitatively change our results for numerical abundance because zooplankton numerical abundance is on average three orders of magnitude less than that of phytoplankton. Without multiplying zooplankton numerical abundance by 6, the biomass abundance of phytoplankton would exceed that of zooplankton. After the multiplication by 6, zooplankton biomass abundance is roughly equal to phytoplankton biomass abundance. For the other relationships involving abundance, such as the abundance-body size allometry, there are no qualitative changes. Slopes and intercepts are quantitatively, but not qualitatively, affected by adjusting the abundance of zooplankton, because the adjustment factor is small compared to the range in abundance in the community.

Sixth, despite attempts to make highly reliable measurements (see Section IV.B), uncertainty remains over how well the estimates of body size and abundance reflect the mean values of temporally variable quantities. We tried to analyze the effect of data variability on the relationship between body size and numerical abundance (Fig. 5). To simulate variation in the data, we randomly and independently perturbed the 1984 estimates of body mass and numerical abundance of each species simultaneously. The perturbed values were drawn from a log-normal distribution with a mean equal to the observed log value and a standard deviation of 0.25 (approximately 95% of the perturbed values will be found within an order of magnitude of the observed values). The variance of the slopes of the linear regression of log numerical abundance as a function of log body mass in 10,000 replicates was 3.38×10^{-4} (CV = 2.25%, mean ± 95% CI: -0.8187 ± 0.00036). This pattern is rather robust to moderate random variation, given the particular model assumptions used here. A different distribution of perturbed values or allowing correlated variation in the data (so that a larger than observed value for body size is associated with a larger than observed abundance) may alter the conclusion. The approach outlined here may in principle be used to analyze effects of variation on other patterns. The primary reason for the robustness of the relationship to variation in the data is the large range of body size and abundance of the species in Tuesday Lake. Body mass and numerical abundance span approximately 12 and 10 orders of magnitude, respectively. Variation within an order of magnitude at the species level will have small effects at the level of the community.

Seventh, this study has a sample size consisting of just one ecosystem, Tuesday Lake. We do not know which relationships described here are unique to Tuesday Lake and which hold in other communities. It would be highly desirable to carry out parallel analyses to test the generality of the patterns described here using data for several different ecosystems, for example, above-ground terrestrial, pedologic, pelagic, and benthic marine, estuarine, and limnic ecosystems (Chase, 2000; Jan Bengtsson, personal communication, 2002). While our present sample size of one ecosystem is not a persuasive basis for generalization, it represents a first step. Our example is intended as a challenge to experts who know the data on other ecosystems.

In summary, the extent to which the data limitations mentioned above affect the patterns reported here is unknown. We therefore call for improved data from similar and different ecosystems to corroborate or challenge the relationships reported here.

VIII. CONCLUSIONS

This study has analyzed the relationships among species abundance, body size, and the food web in the pelagic community of Tuesday Lake. This analysis illustrates a new integrated approach, using a new data structure for the description of ecological communities. A traditional food web (Camerano, 1880) is a directed graph in which each node is associated with a species' name and each arrow (link or directed edge) indicates a flow of nutrients from a resource species to a consumer species. The new data structure introduced here associates with each node a vector of quantitative attributes of the named species (Fig. 1). In this study, the attributes are body size and abundance (numerical and biomass). Since the relationship among these community characteristics affects many other aspects of the community, awareness of these connections is needed for a better understanding of the ecological constraints acting on species assemblages.

In the famous tale of the blind men and the elephant, the blind men cannot agree because they are experiencing different parts of the strange animal. The diverse patterns analyzed here are like the trunk, ears, legs, and tail of the elephant: they all follow from the food web and the body size and abundance of the species in the community (Cohen, 1991). A clear vision of these three features, and their connectedness, gives a more comprehensive picture of the ecological elephant (Table 1). We have identified some relationships that rarely have been analyzed for entire communities before (e.g., trophic generality and vulnerability with respect to trophic height, body size, and abundance of the species within a food web; abundance-body size allometry; predator-prey abundance allometry). The relationship between the trophic height of a species and its body size or abundance has, to our knowledge, not been analyzed quantitatively in a community before. Many previously reported patterns have been confirmed. Furthermore, body size and abundance are often claimed to be allometrically related, but the exact form of the relationship is disputed. Whole communities have rarely been analyzed before. New insights have been gained from a knowledge of the trophic relations among the species. Many of the relations appear to be very robust to a major perturbation (Section VI). If this finding for Tuesday Lake applies generally, then communities may have properties that are fairly consistent and predictable.

Different fields of ecology have focused on different sets of the bivariate relationships in Table 1. For example, the biomass abundance-body size spectrum (Section V.B.2.b) has mainly been studied by limnologists (Kerr and Dickie, 2001), while studies of rank-abundance and predator-prey body size relationships mainly are confined to the field of terrestrial ecology.

Integration of the relationships as suggested here could bring these fields together.

At least three major tasks remain: (1) to test the generality of the present findings by analyzing comparable or better data on other communities, including temporal and spatial variation and heterotrophic microorganisms and parasites; (2) to explain whatever patterns consistently emerge with persuasive quantitative theory; and (3) to extend and apply the data structure introduced here, which is formally a directed graph with vector-labeled nodes.

Here are some examples of how the data structure introduced here could be extended and applied. If dates and places of observation were added to data on body size and abundance associated with each node of a food web, a dynamic, spatially explicit description would become possible. If each node also had an associated Leslie matrix, in which fertility coefficients depended on the abundance of species consumed by the nodal species, and in which the survival coefficients depended on the abundance of the species that consume the nodal species, then dynamic modeling of age- or stage-structured populations (Caswell, 2001) could be integrated with dynamic food web modeling. Such modeling would promote the general integration of population biology and community ecology. If chemical compositions of each species were added to the vector of attributes (Sterner et al., 1996; Sterner and Elser 2002) and if all coefficients of the Leslie matrix also took explicit account of abiotic environmental variables (such as chemical concentrations of nutrients and toxins), then population biology and community ecology could move toward an integration with biogeochemistry. Additional future prospects are suggested by Brown and Gillooly (2003).

A vector of attributes could be associated with each edge to quantify the flows of energy (Baird and Ulanowicz, 1989) and materials (nutrients and toxins and inert matter), including averages and measures of temporal and/or spatial variation. Empirically estimated energy flows in a community could be compared with the flows predicted by the mortality rates derived from the Leslie matrices, dynamically and at steady state.

The new data structure illustrated in this study and future extensions hold the potential to embed studies of food web structure in a general framework for analyzing communities and ecosystems.

ACKNOWLEDGMENTS

Valuable comments were provided by Jan Bengtsson, Tim Blackburn, Hal Caswell, Bo Ebenman, Ursula Gaedke, Annie Jonsson, Daniel C. Reuman and David Tilman. J.E.C. and T.J. thank Kathe Rogerson for assistance. J.E.C. thanks the U.S. National Science Foundation for grants

BSR92-07293 and DEB-9981552 and Mr. and Mrs. William T. Golden for hospitality during this work. S.R.C. is grateful to S.I. Dodson and the late T.M. Frost for discussions of zooplankton diets, and to J.F. Kitchell and J.R. Hodgson for collaboration on the field experiments. The field work was supported by the National Science Foundation and the A.W. Mellon Foundation. This collaboration between J.E.C. and S.R.C. began in 1988 at a workshop in Santa Fe sponsored by the National Science Foundation. Table 1 and Figs. 1A, 2A,C, 4A and 7A,C, which are reprinted from Cohen *et al.* (2003), are © 2003 by National Academy of Sciences.

APPENDICES

Appendix 1A   Species in Tuesday Lake in 1984

Id #	Species name (category)	*BM*	*NA*	*TS*	*TH*
1	*Nostoc* sp. (P)	7.97×10^{-13}	2.00×10^{6}	1	1
2	*Arthrodesmus* sp. (P)	1.52×10^{-12}	4.90×10^{7}	2	1
3	*Asterionella formosa* (P)	1.12×10^{-12}	5.00×10^{6}		
4	*Cryptomonas* sp. 1 (P)	2.03×10^{-13}	6.40×10^{7}	3	1
5	*Cryptomonas* sp. 2 (P)	1.51×10^{-12}	2.80×10^{7}	4	1
6	*Chroococcus dispersus* (P)	2.39×10^{-13}	2.00×10^{7}	3	1
7	*Closteriopsis longissimus* (P)	2.37×10^{-13}	1.00×10^{8}	5	
8	*Chrysosphaerella longispina* (P)	8.31×10^{-10}	4.00×10^{6}		
9	*Dinobryon bavaricum*[†] (P)	2.44×10^{-12}	3.00×10^{7}	6	1
10	*Dinobryon cylindricum*[†] (P)	1.57×10^{-12}	3.00×10^{6}	1	1
11	*Dactylococcopsis fascicularis* (P)	1.32×10^{-13}	4.60×10^{7}	1	1
12	*Diceras* sp. (P)	1.53×10^{-13}	1.40×10^{7}		
13	*Dictyosphaerium pulchellum* (P)	5.07×10^{-13}	1.30×10^{7}	4	1
14	*Dinobryon sertularia* (P)	1.52×10^{-11}	2.00×10^{6}	7	1
15	*Dinobryon sociale* (P)	6.41×10^{-13}	2.80×10^{7}	4	1
16	*Glenodinium quadridens* (P)	7.54×10^{-12}	6.70×10^{7}	8	1
17	*Microcystis aeruginosa*[‡] (P)	1.62×10^{-11}	1.30×10^{7}	6	1
18	*Mallomonas* sp. 1 (P)	1.03×10^{-12}	1.90×10^{7}	7	1
19	*Mallomonas* sp. 2 (P)	1.41×10^{-12}	2.27×10^{7}	2	1
20	Unclassified *flagellates* (P)	3.46×10^{-13}	1.88×10^{9}	3	1
21	*Peridinium limbatum* (P)	6.46×10^{-11}	1.70×10^{7}	6	1
22	*Peridinium cinctum* (P)	4.06×10^{-11}	8.00×10^{6}	7	1
23	*Peridinium pulsillum* (P)	1.58×10^{-12}	1.23×10^{8}	4	1
24	*Peridinium wisconsinense* (P)	3.56×10^{-11}	1.40×10^{7}	6	1
25	*Chromulina* sp. (P)	3.03×10^{-14}	1.49×10^{8}	3	1
26	*Rhizosolenia* sp. (P)	6.86×10^{-13}	5.60×10^{7}		
27	*Selenastrum minutum* (P)	2.72×10^{-13}	2.00×10^{8}	3	1
28	*Spinocosmarium* sp. (P)	3.71×10^{-12}	2.00×10^{6}		
29	*Staurastrum* sp. (P)	4.30×10^{-12}	9.00×10^{6}		
30	*Synedra* sp. (P)	9.18×10^{-11}	1.00×10^{6}	6	1
31	*Trachelomonas* sp. (P)	1.75×10^{-13}	2.22×10^{8}	3	1

(*Continued*)

Appendix 1A *(Continued)*

Id #	Species name (category)	BM	NA	TS	TH
32	*Ascomorpha eucadis*[¥] (Z)	1.40×10^{-10}	2.30×10^3	9	2
33	*Synchaeta sp.*[¥] (Z)	9.50×10^{-10}	5.00×10^3	9	2
34	*Bosmina longirostris* (Z)	1.55×10^{-9}	2.59×10^4	10	2
35	*Conochilus* (solitary) (Z)	3.50×10^{-11}	6.00×10^3	11	2
36	*Cyclops varians rubellus* (Z)	2.04×10^{-8}	1.30×10^3	12	3
37	*Diaphanosoma leuchtenbergianum* (Z)	2.24×10^{-9}	2.40×10^3	13	2
38	*Daphnia pulex* (Z)	5.80×10^{-8}	3.00×10^2	14	2.42
39	*Filinia longispina* (Z)	1.80×10^{-10}	4.00×10^2	15	2
40	*Conochiloides dossuarius* (Z)	1.60×10^{-10}	3.91×10^4	11	2
41	*Gastropus stylifer* (Z)	1.35×10^{-10}	5.90×10^3	15	2
42	*Holopedium gibberum* (Z)	8.75×10^{-8}	1.00×10^2	16	2
43	*Kellicottia sp.*[∈] (Z)	2.00×10^{-11}	4.26×10^4	15	2
44	*Keratella cochlearis*[¥] (Z)	1.00×10^{-11}	7.11×10^4	9	2
45	*Keratella testudo* (Z)	1.00×10^{-11}	1.00×10^3	15	2
46	*Leptodiaptomus siciloides* (Z)	8.80×10^{-9}	4.00×10^2	17	2
47	*Orthocyclops modestus* (Z)	2.29×10^{-8}	3.80×10^3	18	4
48	*Ploesoma sp.* (Z)	1.05×10^{-10}	9.30×10^3	15	2
49	*Polyarthra vulgaris* (Z)	4.65×10^{-10}	6.26×10^4	15	2
50	*Trichocerca multicrinis* (Z)	2.50×10^{-10}	7.80×10^3	15	2
51	*Trichocerca cylindrica* (Z)	3.80×10^{-10}	1.36×10^4	15	2
52	*Tropocyclops prasinus* (Z)	6.85×10^{-9}	8.20×10^3	12	3.5
53	*Chaoborus punctipennis* (Z)	3.00×10^{-7}	2.00×10^3	19	4.40
54	*Phoxinus eos* (F)	1.01×10^{-3}	1.97×10^0	20	4.97
55	*Phoxinus neogaeus* (F)	1.17×10^{-3}	1.33×10^{-1}	20	4.97
56	*Umbra limi* (F)	1.29×10^{-3}	1.32×10^{-1}	21	5.64

[†]: eat bacteria.
[‡]: Can be egested by 37, but survives with nutrients absorbed from predator's digestive tract.
[¥]: Is killed by 37, but not consumed.
[∈]: *K. bostoniensis* + *K. longispina*.
Category P: phytoplankton, Z: zooplankton, and F: fish. *BM*: Body mass (kg), *NA*: Numerical abundance (individuals/m³), *TS*: Trophic species number (Appendix 1B), *TH*: Trophic height of species (see text). All *NA* values for zooplankton species only should be multiplied by 6 to convert them to concentrations in the epilimnion, as in all statistical calculations reported here. Missing values indicate isolated species.

Appendix 1B Predation matrix of the trophic species web of Tuesday Lake in 1984 (isolated species not included). Biological species with identical prey and identical predators are aggregated into trophic species according to *TS* column of Appendix 1A

	1	2	3	4	5	6	7	8	9	10	11	12	13	14	15	16	17	18	19	20	21
1	0	0	0	0	0	0	0	0	0	0	0	0	1	1	0	1	1	0	0	0	0
2	0	0	0	0	0	0	0	0	0	0	0	0	0	1	0	1	1	0	0	0	0
3	0	0	0	0	0	0	0	0	1	1	1	0	1	1	1	1	1	0	0	0	0
4	0	0	0	0	0	0	0	0	0	1	0	0	1	1	0	1	1	0	0	0	0
5	0	0	0	0	0	0	0	0	0	0	0	1	0	0	1	0	0	0	0	0	0
6	0	0	0	0	0	0	0	0	0	0	0	0	0	0	1	0	0	0	0	0	0
7	0	0	0	0	0	0	0	0	0	0	0	0	0	0	1	0	1	0	0	0	0
8	0	0	0	0	0	0	0	0	0	1	0	0	1	1	0	1	1	0	1	0	0
9	0	0	0	0	0	0	0	0	0	0	0	1	0	1	0	0	0	1	1	0	0
10	0	0	0	0	0	0	0	0	0	0	0	0	0	0	0	0	0	0	1	1	1
11	0	0	0	0	0	0	0	0	0	0	0	1	0	0	0	0	0	1	1	0	0
12	0	0	0	0	0	0	0	0	0	0	0	1	0	0	0	0	0	1	1	1	1
13	0	0	0	0	0	0	0	0	0	0	0	0	0	0	0	0	0	0	1	1	1
14	0	0	0	0	0	0	0	0	0	0	0	0	0	0	0	0	0	0	1	1	1
15	0	0	0	0	0	0	0	0	0	0	0	1	0	0	0	0	0	1	1	0	0
16	0	0	0	0	0	0	0	0	0	0	0	0	0	0	0	0	0	0	0	1	1
17	0	0	0	0	0	0	0	0	0	0	0	1	0	0	0	0	0	1	1	1	1
18	0	0	0	0	0	0	0	0	0	0	0	0	0	0	0	0	0	1	1	1	1
19	0	0	0	0	0	0	0	0	0	0	0	0	0	0	0	0	0	0	1	1	1
20	0	0	0	0	0	0	0	0	0	0	0	0	0	0	0	0	0	0	0	0	1
21	0	0	0	0	0	0	0	0	0	0	0	0	0	0	0	0	0	0	0	0	1

Appendix 2A Species in Tuesday Lake in 1986

Id #	Species name (category)	*BM*	*NA*	*TS*	*TH*
1	*Anabaena circinalis* (P)	1.91×10^{-13}	6.00×10^6		
2	*Ankyra judayi* (P)	1.53×10^{-13}	1.30×10^7	1	1
3	*Cryptomonas* sp. 1 (P)	2.85×10^{-13}	3.30×10^7	2	1
4	*Cryptomonas* sp. 3 (P)	6.72×10^{-13}	1.80×10^7	3	1
5	*Cryptomonas* sp. 4 (P)	1.64×10^{-12}	2.80×10^7	3	1
6	*Chroococcus dispersus* (P)	2.39×10^{-13}	5.00×10^6	2	1
7	*Chroococcus limneticus* (P)	1.31×10^{-12}	1.60×10^7	2	1
8	*Cosmarium* sp. (P)	3.71×10^{-12}	1.00×10^6	3	1
9	*Closteriopsis longissimus* (P)	1.98×10^{-13}	1.00×10^6	4	1
10	*Chrysosphaerella longispina* (P)	4.40×10^{-11}	1.00×10^6		
11	*Dinobryon bavaricum* (P)	5.29×10^{-12}	8.00×10^6	5	1
12	*Dinobryon cylindricum* (P)	4.48×10^{-12}	6.70×10^7	5	1
13	*Dactylococcopsis fascicularis* (P)	1.32×10^{-13}	1.00×10^6	3	1
14	*Diceras* sp. (P)	1.53×10^{-13}	1.00×10^6		
15	*Dictyosphaerium pulchellum* (P)	4.07×10^{-13}	1.00×10^6	3	1
16	*Dinobryon sertularia* (P)	6.28×10^{-12}	2.00×10^6	3	1
17	*Sphaerocystis schroeteri*[†] (P)	1.08×10^{-11}	2.00×10^6	3	1
18	*Gloeocystis* sp.[‡] (P)	9.46×10^{-11}	5.00×10^6	5	1

(*Continued*)

Appendix 2A   (*Continued*)

Id #	Species name (category)	BM	NA	TS	TH
19	*Glenodinium pulvisculus* (P)	5.20×10^{-12}	8.00×10^{6}	3	1
20	*Microcystis aeruginosa*[‡] (P)	1.62×10^{-11}	2.00×10^{6}	5	1
21	*Mallomonas*-spiny sp. 1 (P)	2.22×10^{-12}	2.10×10^{7}		
22	*Mallomonas*-spiny sp. 2 (P)	2.22×10^{-12}	2.60×10^{7}		
23	unclassified *microflagellates* (P)	1.02×10^{-13}	1.26×10^{8}	2	1
24	*Oocystis* sp. 1 (P)	3.86×10^{-12}	2.40×10^{7}	3	1
25	*Oocystis* sp. 2 (P)	6.32×10^{-12}	3.00×10^{6}	3	1
26	*Oscillatoria* sp. (P)	1.61×10^{-12}	6.00×10^{6}	6	1
27	*Peridinium limbatum* (P)	6.46×10^{-11}	1.00×10^{6}	5	1
28	*Peridinium pulsillum* (P)	1.58×10^{-12}	1.00×10^{6}	3	1
29	*Quadrigula lacustris* (P)	7.13×10^{-12}	1.03×10^{8}	5	1
30	*Quadrigula* sp. 2 (P)	9.48×10^{-13}	1.10×10^{7}	5	1
31	*Chromulina* sp. (P)	3.03×10^{-14}	2.09×10^{8}	2	1
32	*Schroederia setigera* (P)	6.37×10^{-13}	8.90×10^{7}	3	1
33	*Selenastrum minutum* (P)	2.72×10^{-13}	1.10×10^{7}	2	1
34	*Synedra* sp. (P)	3.62×10^{-13}	2.00×10^{6}	5	1
35	*Synura* sp. (P)	5.07×10^{-12}	1.00×10^{6}		
36	*Ascomorpha eucadis* (Z)	4.00×10^{-10}	3.50×10^{3}	7	2
37	*Conochilus* (colonial) (Z)	1.46×10^{-8}	7.00×10^{2}	8	2
38	*Conochiloides dossuarius* (Z)	1.60×10^{-10}	3.00×10^{2}	9	2
39	*Cyclops varians rubellus* (Z)	2.44×10^{-8}	4.00×10^{2}	10	3
40	*Diaptomus oregonensis* (Z)	1.44×10^{-8}	1.00×10^{2}	11	2
41	*Daphnia pulex* (Z)	4.56×10^{-8}	2.60×10^{3}	12	2.39
42	*Daphnia rosea* (Z)	1.36×10^{-8}	4.00×10^{2}	13	2.47
43	*Gastropus hyptopus* (Z)	1.35×10^{-10}	3.00×10^{2}	14	2
44	*Gastropus stylifer* (Z)	1.00×10^{-10}	1.90×10^{3}	15	2
45	*Holopedium gibberum* (Z)	4.89×10^{-8}	7.00×10^{2}	16	2
46	*Kellicottia bostoniensis* (Z)	2.00×10^{-11}	5.30×10^{3}	14	2
47	*Kellicottia longispina* (Z)	4.50×10^{-11}	5.00×10^{2}	14	2
48	*Keratella cochlearis* (Z)	1.00×10^{-11}	8.80×10^{3}	17	2
49	*Keratella testudo* (Z)	1.50×10^{-11}	1.16×10^{4}	14	2
50	*Orthocyclops modestus* (Z)	3.22×10^{-8}	1.00×10^{2}	10	3.5
51	*Polyarthra vulgaris* (Z)	2.60×10^{-10}	1.26×10^{4}	14	2
52	*Synchaeta* sp. (Z)	3.70×10^{-10}	4.90×10^{3}	17	2
53	*Trichocerca cylindrica* (Z)	5.90×10^{-10}	3.70×10^{3}	14	2
54	*Trichocerca multicrinis* (Z)	7.00×10^{-11}	7.00×10^{2}	14	2
55	*Tropocyclops prasinus* (Z)	8.95×10^{-9}	2.00×10^{2}	18	4.02
56	*Chaoborus punctipennis* (Z)	2.10×10^{-7}	2.00×10^{3}	19	4.39
57	*Micropterus salmoides* (F)	1.95×10^{-1}	3.40×10^{-2}	20	5.24

[†]: Can be egested by its predators, but survives with nutrients absorbed from predators' digestive tract.
[‡]: Can be egested by 41, but survives with nutrients absorbed from predator's digestive tract.
Category P: Phytoplankton, Z: Zooplankton and F: Fish. *BM* Body mass (kg), NA: Numerical abundance (individuals/m³), *TS*: Trophic species number (Appendix 2B), *TH*: Trophic height of species (see text). All *NA* values for zooplankton species only should be multiplied by 6 to convert them to concentrations in the epilimnion, as in all statistical calculations reported here.
Missing values indicate isolated species.

Appendix 2B Predation matrix of the trophic species web of Tuesday Lake in 1986 (isolated species not included). Biological species with identical prey and identical predators are aggregated into trophic species according to *TS* column of Appendix 2A

	1	2	3	4	5	6	7	8	9	10	11	12	13	14	15	16	17	18	19	20
1	0	0	0	0	0	0	0	0	0	0	0	0	1	0	1	0	0	0	0	0
2	0	0	0	0	0	0	1	1	1	0	1	1	1	1	1	1	1	0	0	0
3	0	0	0	0	0	0	0	0	0	0	1	1	1	0	0	1	0	0	0	0
4	0	0	0	0	0	0	0	1	1	0	0	1	0	0	0	0	0	0	0	0
5	0	0	0	0	0	0	0	0	0	0	0	1	0	0	0	0	0	0	0	0
6	0	0	0	0	0	0	0	0	0	0	0	1	1	0	0	0	0	0	0	0
7	0	0	0	0	0	0	0	0	0	0	1	0	1	1	0	0	0	0	0	1	0
8	0	0	0	0	0	0	0	0	0	0	0	0	0	0	0	0	0	0	0	0
9	0	0	0	0	0	0	0	0	0	1	0	0	0	0	0	0	0	1	1	0
10	0	0	0	0	0	0	0	0	0	1	0	0	0	0	0	0	0	1	1	0
11	0	0	0	0	0	0	0	0	0	1	0	0	0	0	0	0	0	1	1	0
12	0	0	0	0	0	0	0	0	0	0	0	0	0	0	0	0	0	0	1	1
13	0	0	0	0	0	0	0	0	0	0	0	0	0	0	0	0	0	0	1	0
14	0	0	0	0	0	0	0	0	0	1	0	0	0	0	0	0	0	1	1	0
15	0	0	0	0	0	0	0	0	0	1	0	0	0	0	0	0	0	1	1	0
16	0	0	0	0	0	0	0	0	0	0	0	0	0	0	0	0	0	0	0	1
17	0	0	0	0	0	0	0	0	0	1	0	1	1	0	0	0	0	1	1	0
18	0	0	0	0	0	0	0	0	0	1	0	0	0	0	0	0	0	1	1	0
19	0	0	0	0	0	0	0	0	0	0	0	0	0	0	0	0	0	0	1	1
20	0	0	0	0	0	0	0	0	0	0	0	0	0	0	0	0	0	0	0	1

REFERENCES

Baird, D. and Ulanowicz, R.E. (1989) The seasonal dynamics of the Chesapeake Bay ecosystem. *Ecol. Monogr.* **59**, 329–364.

Bergquist, A.M. (1985) *Effects of herbivory on phytoplankton community composition, size structure and primary production.* Ph.D. dissertation, University of Notre Dame, Notre Dame, Indiana, USA.

Bergquist, A.M. and Carpenter, S.R. (1986) Limnetic herbivory: Effects on phytoplankton populations and primary production. *Ecology.* **67**, 1351–1360.

Bergquist, A.M., Carpenter, S.R. and Latino, J.C. (1985) Shifts in phytoplankton size structure and community composition during grazing by contrasting zooplankton assemblages. *Limnol. Oceanogr.* **30**, 1037–1045.

Blackburn, T.M. and Gaston, K.J. (1994) Animal body size distributions: Patterns, mechanisms and implications. *Trends Ecol. Evol.* **9**, 471–474.

Blackburn, T.M. and Gaston, K.J. (1997) A critical assessment of the form of the interspecific relationship between abundance and body size in animals. *J. Anim. Ecol.* **66**, 233–249.

Blackburn, T.M. and Gaston, K.J. (1999) The relationship between animal abundance and body size: A review of the mechanisms. *Adv. Ecol. Res.* **28**, 181–210.

Blackburn, T.M., Lawton, J.H. and Pimm, S.L. (1993) Non-metabolic explanations for the relationship between body size and animal abundance. *J. Anim. Ecol.* **62**, 694–702.

Boers, P., van Ballegooijen, L. and Uunk, J. (1991) Changes in phosphorus cycling in a shallow lake due to food web manipulations. *Freshwater Biol.* **25**, 9–20.

Borgman, U. (1987) Models on the slope of, and biomass flow up, the biomass size spectrum. *Can. J. Fish. Aquat. Sci.* **44**, 136–140.

Brown, J.H. and Gillooly, J.F. (2003) Ecological food webs: High-quality data facilitate theoretical unification. *Proc. Natl. Acad. Sci.* **100**, 1467–1468.

Brown, J.H. and Maurer, B.A. (1987) Evolution of species assemblages: Effects of energetic constraints and species dynamics on the diversification of the North American avifauna. *Am. Nat.* **130**, 1–17.

Brown, J.H. and Nicoletto, P.F. (1991) Spatial scaling of species composition: Body masses of North American land mammals. *Am. Nat.* **138**, 1478–1512.

Calder, W.A. (1984) *Size, function and life history.* Harvard University Press, Cambridge, MA, USA.

Camerano, L. (1880) Dell'equilibrio dei viventi mercè la reciproca distruzione. *Atti della Reale Accademia delle Scienze di Torino* **15**, 393–414. (Translation by Claudia M. Jacobi, edited by J.E. Cohen, 1994.) On the equilibrium of living beings due to reciprocal destruction. In: *Frontiers in Theoretical Biology* (Ed. by S.A. Levin), *Biomathematics* 100, pp. 360–380. Springer-Verlag, New York, USA.

Carbone, C. and Gittleman, J.L. (2002) A common rule for the scaling of carnivore density. *Science* **295**, 2273–2276.

Carpenter, S.R. and Kitchell, J.F. (1988) Consumer control of lake productivity. *Bioscience* **38**, 764–769.

Carpenter, S.R. and Kitchell, J.F. (1993a) *The trophic cascade in lakes.* Cambridge University Press, Cambridge, UK.

Carpenter, S.R. and Kitchell, J.F. (1993b) Experimental lakes, manipulations and measurements. In: *The trophic cascade in lakes* (Ed. by S.R. Carpenter and J.F. Kitchell), pp. 15–25. Cambridge University Press, Cambridge, UK.

Carpenter, S.R., Kitchell, J.F. and Hodgson, J.R. (1985) Cascading trophic interactions and lake productivity. *Bioscience* **35**, 634–639.

Carpenter, S.R., Kraft, C.E., Wright, R., He, X., Soranno, P. and Hodgson, J.R. (1992) Resilience and resistance of a lake phosphorous cycle before and after food web manipulation. *Am. Nat.* **140**, 781–798.

Caswell, H. (1988) Theory and models in ecology: A different perspective. *Ecol. Mod.* **43**, 33–44.

Caswell, H. (2001) *Matrix population models: Construction, analysis, and interpretation.* Sinauer Associates, Sunderland MA, USA.

Charnov, E.L. (1993) *Life history invariants: Some explorations of symmetry in evolutionary ecology.* Oxford University Press, Oxford, UK.

Chase, J.M. (2000) Are there real differences among aquatic and terrestrial food webs? *Trends Ecol. Evol.* **15**, 408–412.

Closs, G.P. and Lake, P.S. (1994) Spatial and temporal variation in the structure of an intermittent-stream food web. *Ecol. Monogr.* **64**, 1–21.

Cochran, P.A., Lodge, D.M., Hodgson, J.R. and Knapik, P.G. (1988) Diets of syntopic finescale dace, *Phoxinus neogaeus*, and northern redbelly dace, *Phoxinus eos*: A reflection of trophic morphology. *Env. Biol. Fish.* **22**, 235–240.

Cohen, J.E. (1966) *A model of simple competition.* Harvard University Press, Cambridge, MA, USA.

Cohen, J.E. (1968) Alternate derivations of a species-abundance relation. *Am. Nat.* **102**, 165–172.

Cohen, J.E. (1989) Food webs and community structure. In: *Perspectives in Theoretical Ecology* (Ed. by J. Roughgarden, R.M. May and S.A. Levin), Princeton University Press, Princeton, USA.

Cohen, J.E. (1990) A stochastic theory of community food webs. VI. Heterogeneous alternatives to the cascade model. *Theor. Popul. Biol.* **37**, 55–90.

Cohen, J.E. (1991) Food webs as a focus for unifying ecological theory. *Ecol. Intl. Bull.* **19**, 1–13.

Cohen, J.E., Briand, F. and Newman, C.M. (1986) A stochastic theory of community food webs III. Predicted and observed lengths of food chains. *Proc. R. Soc. Lond. B.* **228**, 317–353.

Cohen, J.E., Briand, F. and Newman, C.M. (1990) *Community food webs: Data and theory.* Springer-Verlag, Berlin, Germany.

Cohen, J.E. and Carpenter, Stephen, R. (2005) Species' average body mass and numerical abundance in a community food web: Statistical questions in estimating the relationship. In: *Dynamic Food Webs: Multispecies Assemblages, Ecosystem Development and Environmental Change* (Ed. by Peter C. de Ruiter, Volkmar Wolters and John C. Moore), Elsevier, Amsterdam (in press).

Cohen, J.E. and Newman, C.M. (1991) Community area and food-chain length: Theoretical predictions. *Am. Nat.* **138**, 1542–1554.

Cohen, J.E. and Palka, Z.J. (1990) A stochastic theory of community food webs: V. Intervality and triangularity in the trophic niche overlap graph. *Am. Nat.* **135**, 435–463.

Cohen, J.E., Jonsson, T. and Carpenter, S.R. (2003) Ecological community description using the food web, species abundance, and body size. *Proc. Nat. Acad. Sci.* **100**, 1781–1786.

Cohen, J.E., Pimm, S.L., Yodzis, P. and Saldaña, J. (1993) Body sizes of animal predators and animal prey in food webs. *J. Anim. Ecol.* **62**, 67–78.

Cousins, S. (1987) The decline of the trophic level concept. *Trends Ecol. Evol.* **2**, 312–316.

Cousins, S.H. (1996) Food webs: From the Lindeman paradigm to a taxonomic general theory of ecology. In: *Food webs: Integration of patterns and dynamics* (Ed. by G. A. Polis and K.O. Winemiller), pp. 243–251. Chapman & Hall, New York, USA.

Cyr, H., Downing, J.A. and Peters, R.H. (1997a) Density-body size relationships in local aquatic communities. *Oikos* **79**, 333–346.

Cyr, H., Peters, R.H. and Downing, J.A. (1997b) Population density and community size structure: Comparison of aquatic and terrestrial systems. *Oikos* **80**, 139–149.

D'Agostino, R.B. and Pearson, E.S. (1973) Tests of departure from normality. Empirical results for the distribution of b_2 and $(b_1)^{1/2}$. *Biometrika.* **60**, 613–622.

Damuth, J. (1981) Population density and body size in mammals. *Nature* **290**, 699–700.

Darwin, C. and Wallace, R.A. (1858) On the tendency of species to form varieties; and on the perpetuation of varieties and species by means of natural selection. *Proc. Linnean Soc. Lond.* **3**, 53–62.

de Ruiter, P.C., Neutel, A.-M. and Moore, J.C. (1994) Modelling food webs and nutrient cycling in agro-ecosystems. *Trends Ecol. Evol.* **9**, 378–383.

de Ruiter, P.C., Neutel, A.-M. and Moore, J.C. (1995) Energetics, patterns of interaction strengths, and stability in real ecosystems. *Science* **269**, 1257–1260.

Del Giorgio, P.A. and Gasol, J.M. (1995) Biomass distribution in freshwater plankton communities. *Am. Nat.* **146**, 135–152.

Dini, M.L., Soranno, P.A., Scheuerell, M.D. and Carpenter, S.R. (1993) Effects of predators and food supply on diel vertical migration of *Daphnia*. In: *The trophic cascade in lakes* (Ed. by S.R. Carpenter and J.F. Kitchell), pp. 153–171. Cambridge University Press, Cambridge, UK.

Dodds, P.S., Rothman, D.H. and Weitz, J.S. (2001) Re-examination of the "3/4-law" of metabolism. *J. Theor. Biol.* **209**, 9–27.

Elser, J.J., Elser, M.M. and Carpenter, S.R. (1986) Size fractionation of algal chlorophyll, carbon fixation, and phosphatase activity: Relationships with species specific size distributions and zooplankton community structure. *J. Plankton Res.* **8**, 365–383.

Elser, M.M., von Ende, C.N., Soranno, P. and Carpenter, S.R. (1987a) Response to food web manipulations and potential effects on zooplankton communities. *Can. J. Zool.* **65**, 2846–2852.

Elser, J.J., Goff, N.C., MacKay, N.A., St. Amand, A.L., Elser, M.M. and Carpenter, S.R. (1987b) Species specific algal responses to zooplankton: Experimental and field observations in three nutrient limited lakes. *J. Plankton Res.* **9**, 699–717.

Elton, C. (1927) *Animal ecology.* Sidgwick & Jackson, London, UK.

Elton, C. (1933) *The ecology of animals.* Reprinted 1966 by Science Paperbacks and Methuen, London, UK.

Enquist, B.J., Brown, J.H. and West, G.B. (1998) Allometric scaling of plant energetics and population density. *Nature* **395**, 163–165.

Gaedke, U. (1992) The size distribution of plankton biomass in a large lake and its seasonal variability. *Limnol. Oceanogr.* **37**, 1202–1220.

Gasol, J.M., Del Giorgio, P.A. and Duarte, C.M. (1997) Biomass distribution in marine planktonic communities. *Limnol. Oceanogr.* **42**, 1353–1363.

Goldwasser, L. and Roughgarden, J. (1993) Construction and analysis of a large Caribbean food web. *Ecology* **74**, 1216–1233.

Gray, J.S. (1987) Species-abundance patterns. In: *Organization of communities, past and present* (Ed. by J. H. R. Gee and P.S. Giller), pp. 53–67. Blackwell Scientific, Oxford, UK.

Hall, S.J. and Raffaelli, D. (1991) Food-web patterns: Lessons from a species-rich web. *J. Anim. Ecol.* **60**, 823–842.

Harrison, G.W. (1979) Stability under environmental stress: Resistance, resilience, persistence and variability. *Am. Nat.* **113**, 659–669.

Harvey, H.W. (1950) On the production of living matter in the sea off Plymouth. *J. Marine Biol. Assoc. UK* **29**, 97–137.

Harvey, P.H. and Purvis, A. (1999) Understanding the ecological and evolutionary reasons for life history variation: Mammals as a case study. In: *Advanced ecological theory* (Ed. by J. McGlade), pp. 232–248. Blackwell Science Ltd, Oxford, UK.

Havens, K. (1992) Scale and structure in natural food webs. *Science.* **257**, 1107–1109.

Havlicek, T. and Carpenter, S.R. (2001) Pelagic species size distributions in lakes: Are they discontinuous? *Limnol. Oceanogr.* **46**, 1021–1033.

Hemmingsen, A.M. (1960) Energy metabolism as related to body size and respiratory surfaces, and its evolution. *Rep. Steno Memor. Hosp. Nordisk Insul. Lab.* **9**, 6–110.

Heusner, A.A. (1982) Energy metabolism and body size. I. Is the 0.75 mass exponent of Kleiber's equation a statistical artifact? *Resp. Physiol.* **48**, 1–12.

Hill, J.K., Hamer, K.C., Lace, L.A. and Banham, W. M. T. (1995) Effects of selective logging on tropical forest butterflies on Buru, Indonesia. *J. Appl. Ecol.* **32**, 754–760.

Hodgson, J.R., He, X. and Kitchell, J.F. (1993) The fish populations. In: *The trophic cascade in lakes* (Ed. by S.R. Carpenter and J.F. Kitchell), pp. 43–68. Cambridge University Press, Cambridge, UK.

Holling, C.S. (1992) Cross-scale morphology, geometry, and dynamics of ecosystems. *Ecol. Monogr.* **62**, 447–502.

Huxham, M., Raffaelli, D. and Pike, A. (1995) Parasites and food web patterns. *J. Anim. Ecol.* **64**, 168–176.

Janzen, D.H. and Schoener, T.W. (1968) Differences in insect abundance and diversity between wetter and drier sites during a tropical dry season. *Ecology* **49**, 96–110.

Jonsson, T. and Ebenman, B. (1998) Effects of predator-prey body size ratios on the stability of food chains. *J. Theor. Biol.* **193**, 407–417.

Kenny, D. and Loehle, C. (1991) Are food webs randomly connected? *Ecology* **72**, 1794–1799.

Kerfoot, W.C. and DeAngelis, D.L. (1989) Scale-dependent dynamics: Zooplankton and the stability of freshwater food webs. *Trends Ecol. Evol.* **4**, 167–171.

Kerr, S.R. and Dickie, L.M. (2001) *The biomass spectrum: A predator-prey theory of aquatic production.* Columbia University Press, New York.

Kleiber, M. (1932) Body size and metabolism. *Hilgardia* **6**, 315–353.

Kozlowski, J. and Weiner, J. (1997) Interspecific allometries are by-products of body size optimizations. *Am. Nat.* **149**, 352–380.

Lawton, J.H. (1989) Food webs. In: *Ecological concepts: The contribution of ecology to an understanding of the natural world* (Ed. by J.M. Cherrett), pp. 43–78. Blackwell Scientific Publications, Oxford, UK.

Loder, N., Blackburn, T.M. and Gaston, K.J. (1997) The slippery slope: Towards an understanding of the body size frequency distribution. *Oikos* **78**, 195–201.

MacArthur, R.H. (1957) On the relative abundance of bird species. *Proc. Nat. Acad. Sci.* **43**, 293–295.

MacArthur, R.H. (1960) On the relative abundance of species. *Am. Nat.* **94**, 25–36.

MacArthur, R.H. and Wilson, E. (1967) *The theory of island biogeography.* Princeton University Press, Princeton, N.J., USA.

Marcogliese, D.J. and Cone, D.K. (1997) Food webs: A plea for parasites. *Trends Ecol. Evol.* **12**, 320–325.

Marquet, P.A., Navarrete, S.A. and Castilla, J.C. (1990) Scaling population density to body size in rocky intertidal communities. *Science* **250**, 1125–1127.

Martinez, N.D. (1991) Artifacts or attributes: Effects of resolution on the Little Rock Lake food web. *Ecol. Monogr.* **61**, 367–392.

May, R.M. (1972) Will a large complex system be stable? *Nature* **238**, 413–414.

May, R.M. (1975) Patterns of species abundance and diversity. In: *Ecology of species and communities* (Ed. by M. Cody and J.M. Diamond), pp. 81–120. Harvard University Press, Cambridge, MA, USA.

May, R.M. (1986) The search for patterns in the balance of nature: Advances and retreats. *Ecology.* **67**, 1115–1126.

May, R.M. (1989) Levels of organization in ecology. In: *Ecological concepts: The contribution of ecology to an understanding of the natural world* (Ed. by J.M. Cherrett), pp. 339–363. Blackwell Scientific Publications, Oxford, UK.

Memmott, J., Martinez, N.D. and Cohen, J.E. (2000) Predators, parasitoids and pathogens: Species richness, trophic generality and body sizes in a natural food web. *J. Anim. Ecol.* **69**, 1–15.

Menge, B.A. and Sutherland, J.P. (1976) Species diversity gradients: Synthesis of the roles of predation, competition, and temporal heterogeneity. *Am. Nat.* **110**, 351–369.

Mohr, C.O. (1940) Comparative populations of game, fur and other mammals. *Am. Midland Nat.* **24**, 581–584.

Morse, D.R., Stork, N.E. and Lawton, J.H. (1988) Species number, species abundance and body length relationships of arboreal beetles in Bornean lowland rain forest trees. *Ecol. Entomol.* **13**, 25–37.

Muratori, S. and Rinaldi, S. (1992) Low- and high-frequency oscillations in three-dimensional food chain systems. *SIAM J. Appl. Math.* **52**, 1688–1706.

Neubert, M., Blumenshine, S., Duplisea, D., Jonsson, T. and Rashlei, B. (2000) Body size, food web structure, and the cascade model's equiprobability assumption. *Oecologia.* **123**, 241–251.

Neutel, A.-M., Heesterbeek, J. A. P. and de Ruiter, P.C. (2002) Stability in real food webs: Weak links in long loops. *Science* **296**, 1120–1123.

Nummelin, M. (1998) Log-normal distribution of species abundance is not a universal indicator of rain forest disturbance. *J. Appl. Ecol.* **35**, 454–457.

Odum, E.P. (1983) *Basic ecology.* Saunders College Publishing, Philadelphia, USA.

Pace, M.L. (1993) Heterotrophic microbial processes. In: *The trophic cascade in lakes* (Ed. by S.R. Carpenter and J.F. Kitchell), pp. 252–277. Cambridge University Press, Cambridge, UK.

Paine, R.T. (1980) Food webs: Linkage, interaction strength and community infrastructure. *J. Anim. Ecol.* **49**, 667–685.

Persson, L., Diehl, S., Johansson, L., Andersson, G. and Hamrin, S.F. (1992) Trophic interactions in temperate lake ecosystems: A test of food chain theory. *Am. Nat.* **140**, 59–84.

Peters, R.H. (1983) *The ecological implications of body size.* Cambridge University Press, Cambridge, UK.

Peters, R.H. and Raelson, J.V. (1984) Relations between individual size and mammalian population density. *Am. Nat.* **124**, 498–517.

Peters, R.H. and Wassenberg, K. (1983) The effect of body size on animal abundance. *Oecologia.* **60**, 89–96.

Pimm, S.L. (1982) *Food webs.* Chapman and Hall, London, UK.

Polis, G.A. (1991) Complex trophic interactions in deserts: An empirical critique of food-web theory. *Am. Nat.* **138**, 123–155.

Porter, K.G., Pearl, H., Hodson, R., Pace, M., Priscu, J., Rieman, B., Scavia, D. and Stockner, J. (1988) Microbial interactions in lake food webs. In: *Complex interactions in lake communities* (Ed. by S.R. Carpenter), Springer-Verlag, New York, USA.

Purvis, A., Gittleman, J.L., Cowlishaw, G. and Mace, G.M. (2000) Predicting extinction risk in declining species. *Proc. R. Soc. Lon. B.* **267**, 1947–1952.

Raffaelli, D. (2002) From Elton to mathematics and back again. *Science.* **296**, 1035–1037.

Raffaelli, D., Hall, S., Emes, C. and Manly, B. (2000) Constraints on body size distributions: An experimental approach using a small-scale system. *Oecologia.* **122**, 389–398.

Ritchie, M.E. and Olff, H. (1999) Spatial scaling laws yield a synthetic theory of biodiversity. *Nature* **400**, 557–560.

Rodriguez, J. and Mullin, M.M. (1986) Relation between biomass and body weight of plankton in a steady state oceanic ecosystem. *Limnol. Oceanogr.* **31**, 361–370.

Schoener, T.W. (1989) Food webs from the small to the large. *Ecology.* **70**, 1559–1589.

Schoenly, K., Beaver, R.A. and Heumier, T.A. (1991) On the trophic relations of insects: A food-web approach. *Am. Nat.* **137**, 597–638.

Schoenly, K. and Cohen, J.E. (1991) Temporal variation in food web structure: 16 empirical cases. *Ecol. Monogr.* **61**, 267–298.

Schwinghammer, P. (1981) Characteristic size distributions of integral benthic communities. *Can. J. Fish. Aquat. Sci.* **38**, 1255–1263.

Sheldon, R.W., Prakash, A. and Sutcliffe, W.H., Jr. (1972) The size distribution of particles in the ocean. *Limnol. Oceanogr.* **17**, 327–340.

Sheldon, R.W., Sutcliffe, J.H., Jr. and Paranjape, M.A. (1977) Structure of pelagic food chain and relationship between plankton and fish production. *J. Fish. Res. Board Can.* **34**, 2344–2353.

Siemann, E., Tilman, D. and Haarstad, J. (1996) Insect species diversity, abundance and body size relationships. *Nature* **380**, 704–706.

Silva, M. and Downing, J.A. (1995) The allometric scaling of density and body mass: a non-linear relationship for terrestrial mammals. *Am. Nat.* **145**, 704–727.

Solow, A.R. and Beet, A.R. (1998) On lumping species in food webs. *Ecology* **79**, 2013–2018.

St. Amand, A.L. (1990) *Mechanisms controlling metalimnetic communities and the importance of metalimnetic phytoplankton to whole-lake primary productivity.* Ph.D. dissertation, University of Notre Dame, Notre Dame, Indiana.

Sterner, R.W., Elser, J.J., Chrzanowski, T.H., Schampel, J.H. and George, N.B. (1996) Biogeochemistry and trophic ecology: A new food web diagram. In: *Food webs: integration of patterns and dynamics* (Ed. by G.A. Polis and K.O. Winemiller), pp. 72–80. Chapman & Hall, New York, USA.

Sterner, R.W. and Elser, J.J. (2002) *Ecological stoichiometry: The biology of elements from molecules to the biosphere.* Princeton University Press, Princeton and Oxford.

Stockner, J.G. and Porter, K.G. (1988) Microbial food webs in freshwater planktonic ecosystems. In: *Complex interactions in lake communities* (Ed. by S. R. Carpenter), Springer-Verlag, New York, USA.

Tavares-Cromar, A.F. and Williams, D.D. (1996) The importance of temporal resolution in food web analysis: Evidence from a detritus-based stream. *Ecol. Monogr.* **66**, 91–113.

Tokeshi, M. (1993) Species abundance patterns and community structure. *Adv. Ecol. Res.* **24**, 111–186.

Vézina, A.F. (1985) Empirical relationships between predator and prey size among terrestrial vertebrate predators. *Oecologia.* **67**, 555–565.

Vidondo, B., Prairie, Y.T., Blanco, J.M. and Duarte, C.M. (1997) Some aspects of the analysis of size spectra in aquatic ecology. *Limnol. Oceanogr.* **42**, 184–192.

Warren, P.H. (1989) Spatial and temporal variation in the structure of a freshwater food web. *Oikos* **55**, 299–311.

Warren, P.H. (1994) Making connections in food webs. *Trends Ecol. Evol.* **9**, 136–141.

Warren, P.H. and Lawton, J.H. (1987) Invertebrate predator-prey body size relationships: An explanation for upper triangular food webs and patterns in food web structure? *Oecologia* **74**, 231–235.

Watt, A.D. (1998) Measuring disturbance in tropical forests: A critique of the use of species-abundance models and indicator measures in general. *J. Appl. Ecol.* **35**, 469.

Wetzel, R.G. (1983) *Limnology.* Saunders College Publishing, Philadelphia, PA, USA.

Williams, R.J. and Martinez, N.D. (2000) Simple rules yield complex food webs. *Nature* **404**, 180–183.

Witek, Z. and Krajewska-Soltys, A. (1989) Some examples of the epipelagic plankton size structure in high latitude oceans. *J. Plankton Res.* **11**, 1143–1155.

Zar, J.H. (1999) *Biostatistical analysis.* Prentice Hall, Upper Saddle River, NJ, USA.

Quantification and Resolution of a Complex, Size-Structured Food Web

GUY WOODWARD, DOUGIE C. SPEIRS, AND
ALAN G. HILDREW

I. SUMMARY

Previous studies were collated with new data to produce an exceptionally detailed connectance web for Broadstone Stream (UK) that contained 131 species, including the permanent meiofauna (i.e., species that are always passing through a mesh of $500\,\mu m$), and 842 links. Despite its apparent complexity, the structure of this web displayed relatively simple patterns related to body size. For instance, many of the speciose permanent meiofauna were not eaten by the large-bodied, higher predators and thus diet width decreased with increasing predator size. When the permanent meiofauna were excluded from the analysis, however, the opposite was found.

We then assessed body-size relationships within both connectance and quantified webs for the macrofaunal "subweb". The detection of links required considerable sampling effort, especially from the smaller (and intermediate) predators to their prey, suggesting that food web complexity is often seriously underestimated and that this might be further confounded with a potential body-size (and trophic status) bias in less exhaustively sampled webs. Trivariate relationships between body size, abundance, and web structure were apparent, with the majority of links representing consumption of smaller, more abundant prey by larger and rarer predators. In a few instances, this "rule" was broken, largely due to seasonal ontogenetic shifts in body-size distributions. Seasonal changes in resource availability (prey size and abundance) influenced both web complexity and ingestion rates, both of which peaked in summer when generations overlapped, but then declined as prey became scarcer and/or outgrew their potential predators. During summer, connectance in the macrofaunal subweb was high (0.13), and predators ate the equivalent of up to 16.1% of the numerical standing stock per day; by the following spring, however, connectance had halved (0.07) and total consumption had fallen to 5.5% per day. The small predator species ate large numbers of the temporary meiofauna (i.e., mainly insect larvae spending their early instars only in the $<500\,\mu m$ net mesh size class), especially in summer, whereas the large predators ate the same species but exploited larger size classes. However, because consumption of small predators by larger carnivores accounted for about 10% of the production of the latter, macrofaunal-meiofaunal links could provide important "indirect" energy fluxes to the higher trophic levels.

Finally, we produced a quantified food web with the strength of feeding links expressed as annual ingestion (I) of prey by predators divided by the production (P) of that prey. Body size had strong effects on dynamical aspects of the food web (secondary production and interaction strength), in addition to its marked constraining influence on web structure (connectance, diet width, trivariate relationships). We identified multivariate relationships between body size and associated species traits (e.g., P/B ratio, abundance, trophic status) and the strength of interactions within the quantified web. Most links were weak (I/P, <0.01) and interaction strength scaled with both prey and predator body size: the smallest, more "r-selected" species were those most heavily exploited, with virtually all of their annual production consumed. Among the predators, although the small species ate many prey per unit area, this accounted for a relatively trivial amount of prey production. The larger predators, however, had more and stronger interactions with the macroinvertebrate prey assemblage, with the strength of links increasing with predator body size.

II. INTRODUCTION

A. Connectance Food Webs: Early Patterns and Recent Advances

The notions that species within a community are inextricably linked to one another via their ecological interactions, and that the pattern and strength of these links might reveal fundamental properties of a system, particularly in relation to its stability, are two of the earliest ideas in ecology (Elton, 1927; MacArthur, 1955; May, 1973; McCann, 2000). Because trophic links are relatively easy to document—at least when compared with other, more subtle, interactions (e.g., competitive or mutualistic links)—food webs have been the primary focus of research into ecological networks (Cohen et al., 1993a). Nonetheless, constructing realistic empirical food webs is still extremely difficult; most published webs are qualitative "connectance" webs, which are restricted to simple presence/absence data on species (nodes) and feeding links (vertices). Much of the available data are of poor quality: many webs represent only a subset of the wider community food web, taxonomic resolution is inconsistent and biased towards the higher trophic levels, feeding links are often inferred rather than observed, and some published webs contain biological impossibilities (e.g., predators without prey, ducks as basal species; see Hall and Raffaelli, 1993 for a more detailed critique). Attempts have been made to address these shortcomings in several recent studies by quantifying sampling effort (e.g., Woodward and Hildrew, 2001), improving taxonomic resolution (e.g., Schmid-Araya et al., 2002a,b), and including more quantitative information about species (e.g., body-size and abundance data; Cohen et al., 2003) and links (e.g., the flux of matter between species; Benke and Wallace, 1997). Unfortunately, these improvements are often carried out piecemeal in individual studies on different webs, largely due to the inevitable logistic and financial constraints associated with this labor-intensive work and, as a result, there are no webs that are both fully-quantified and highly-resolved for entire communities (McCann, 2000; Berlow et al., 2004). We attempted to address this, at least partially, by exploring the structural properties of one of the most detailed connectance webs currently published, that of Broadstone Stream (Woodward and Hildrew, 2001; Schmid-Araya et al., 2002a), and by subsequently examining relationships between structural patterns and dynamic processes in a quantified subset of the community web, with particular emphasis on the role of body size as a driver of both pattern and process in the food web.

Prior to the mid-1980s, most published connectance webs were very simple, in that they contained few species and/or links (Cohen, 1978; Pimm, 1980, 1982). These webs supported many of the early mathematical models (e.g., May, 1972, 1973) that were in vogue during this period, generally predicting that because the dynamic stability of food webs was inversely related to their complexity, simple webs would predominate in nature and complex webs would be rare (Pimm, 1980). However, many of the early webs were constructed from poor quality data and oversimplified to such an extent that their validity, and hence the theoretical predictions and mathematical models derived from them, have since been questioned and are often roundly criticized (Polis, 1991; Hall and Raffaelli, 1993; McCann, 2000).

Since the mid-1980s, a new catalogue of better quality, data-rich food webs started to emerge, challenging the received wisdom that complexity was the exception rather than the rule (e.g., Hildrew *et al.*, 1985; Warren, 1989; Polis, 1991; Cohen, *et al.*, 1993b, 2003; Closs and Lake, 1994; Tavares-Cromar and Williams, 1996; Benke and Wallace, 1997; Yodzis, 1998; Williams and Martinez, 2000; Woodward and Hildrew, 2001; Schmid-Araya *et al.*, 2002a,b). Many of the apparent discrepancies between the "early and simple" webs and the "recent and complex" webs have now been ascribed to methodological artifacts: the newer webs are far more exhaustively sampled and better resolved taxonomically than many of the earlier webs (Hall and Raffaelli, 1993; Polis, 1994, 1998; McCann, 2000; Woodward and Hildrew, 2002a). For instance, many of the coarse groupings (e.g., algae or meiofauna) that were used previously to lump together supposedly identical "trophic species" (after Sugihara *et al.*, 1997) were now starting to be separated into distinct taxa (e.g., Schmid-Araya *et al.*, 2002a).

Because connectance webs make no distinctions between rare or common species and links, many food web statistics (e.g., connectance, linkage density) are sensitive to resolution and sampling protocol, which are rarely standardized across webs (Cohen *et al.*, 1993a; Hall and Raffaelli, 1993; Martinez *et al.*, 1999). If the guts of most predators are empty, or nearly so (e.g., Woodward and Hildrew, 2001), then links to rare or less-favored prey will be missed, unless sample sizes are very large. However, many studies have used relatively small sample sizes for describing links (Goldwasser and Roughgarden, 1997). For instance, Tavares-Cromar and Williams (1996), Townsend *et al.* (1998) and Thompson and Townsend (1999) analyzed ten guts per taxon on each sampling occasion. Further, sampling effects were potentially autocorrelated with web size, because speciose communities tend to contain more rare species (Tokeshi, 1999), and species are generally more likely to be detected in the community than their food web linkages characterized adequately (e.g., Goldwasser and Roughgarden, 1997; Woodward and Hildrew, 2001). The publication of yield-effort curves for links and species would alleviate this problem, but such curves are rarely shown (Cohen *et al.*, 1993a). Thompson and Townsend (1999) produced curves for species but not for links, whereas Goldwasser and Roughgarden (1997) produced curves for links, but not for individual predator species. Woodward and Hildrew (2001) found that, in a summary web, curves for links varied markedly among species and with trophic status, and that asymptotes were reached only after several hundred guts had been examined. These studies suggest that the small sample sizes used in many of the early studies were unlikely to be sufficient to capture the true complexity of real food webs. None of these studies, however, considered the seasonal variations in yield-effort curves that we address here, which might account for the purported temporal shifts in web structure. Also, little attention has been given to how the strength of a link might be related to the frequency of its detection. Essentially, how important is it to catalog rare links?

The new catalog of highly complex, data-rich empirical webs has driven recent important advances in ecological theory (Polis, 1998; McCann, 2000; Williams and Martinez, 2000), including the recognition of the role of body-size as a structuring force in food webs (Warren, 1996; Cohen *et al.*, 2003; Woodward, *et al.*, in press). A notable and recurrent feature of many of the more recent food webs is the existence of nested hierarchies of dietary niches, such that a given predator's potential diet is effectively a subset of that of the next largest predator (e.g., Cohen *et al.*, 1993b; Woodward and Hildrew, 2002b). Because this generates upper-triangular food web matrices, a central assumption of the cascade model of food web structure (Cohen and Newman, 1985), body size provides one plausible biological explanation for this assumption (Warren and Lawton, 1987; Cohen, 1989).

Subsequent refinements of this type of model have resulted in the development of niche-based food web models (Warren, 1996; Williams and Martinez, 2000). Warren (1996) demonstrated that plausible predictions of web features (e.g., food chain length) could be made when body size was used as a single niche dimension in a model of biological constraints on food web structure. The niche model of Williams and Martinez (2000) also uses a single, general niche dimension (which could represent body size) but the strict hierarchy used in the cascade model is relaxed, so that up to half of the consumer's trophic niche can include species with a niche value higher than itself. This simple model successfully reproduces many of the complex patterns seen in real food webs—such as the prevalence of omnivory—and suggests that community niche space can be collapsed into a single dimension (potentially body size), at least when considering static structural patterns (Williams and Martinez, 2000). Recently, attempts have been made to include greater ecological detail in connectance webs, beyond simple presence/absence data. For instance, Cohen *et al.* (2003) have identified strong trivariate relationships between food web structure, species abundance, and body size. They found that small species were abundant but low in the food web, and that large species were rarer, higher in the food web, and possessed a greater number of links. Despite the fact that these patterns have yet to be examined explicitly in a range of other systems, many of the component univariate or bivariate patterns they report have already been described elsewhere, such as inverse relationships between body size and abundance (Schmid et al., 2000) and upper triangularity (Williams and Martinez, 2000; Woodward and Hildrew, 2001; Schmid-Araya et al., 2002a), suggesting that they might be of general applicability.

B. Quantified Food Webs: From Pattern to Process

One of the major criticisms leveled at connectance webs is that they provide only static representations of the trophic scaffolding within a community, but supply no information about the relative importance of the different links or the dynamic processes (e.g., energy flux; Lotka-Volterra population dynamics) operating within the web (Paine, 1988; Hall and Raffaelli, 1993; Benke and Wallace, 1997). Notwithstanding these objections, connectance webs, despite their many limitations, have undoubtedly provided some invaluable insights into real ecological phenomena; pattern and process are likely to be inextricably linked in real food webs. Increasingly, attempts are being made to unite the static (e.g., Williams and Martinez, 2000) and dynamic (e.g., May, 1972) approaches to food web ecology, and there are suggestions that body size might constrain both the patterning and strength of trophic interactions (e.g., de Ruiter et al., 1995; Emmerson and Raffaelli, 2004).

When considering dynamic processes, the equal weighting of links implicit in connectance webs is unlikely to reflect the true distribution of interaction strengths, which are often highly skewed (Hall and Raffaelli, 1993; McCann, 2000). Species rank-abundance curves typically follow either log-normal or geometric series (Tokeshi, 1999) and there is increasing evidence that, within a web, most links are weak with only a few being strong (e.g., Paine, 1992; de Ruiter et al., 1995; Raffaelli and Hall, 1996; Emmerson and Raffaelli, 2004). Recent models have shown that an abundance of weak links can increase a web's stability (e.g., McCann et al., 1998), forcing ecologists to reassess the long-held paradigm that complex webs are unstable (e.g., May, 1972, 1973; Pimm, 1982). Despite their evident importance in aiding our understanding of community dynamics, interaction strengths are nearly always estimated, rather than measured, and empirical data with which to test the assumptions of models are scarce (but see Emmerson and Raffaelli, 2004).

Attempts to quantify webs can be divided into two broad categories, with the first focusing on population/community dynamics (e.g., Power, 1990; Wootton et al., 1996) and the second on the flux of energy or matter (the "ecosystem approach," e.g., Benke and Wallace, 1997), with virtually none combining the two (but see Hall *et al.*, 2000). The former approach can itself be subdivided into questions about overall community dynamics (e.g., the complexity-stability debate), and those seeking predictions about the dynamics of focal species populations (Berlow *et al.*, 2004). The ecosystem approach has often been used to view food webs from a mass-balance perspective; the recent emergence of stoichiometric analysis, which can be seen as a natural extension of this viewpoint, has led to a closer union with the community approach by recognizing the role of species populations in driving nutrient dynamics (see Elser and Urabe, 1999 for a more detailed review).

There is a small but growing number of quantified (or semi-quantified) food webs in existence (e.g., Power, 1990; Paine, 1992; de Ruiter *et al.*, 1995; Benke and Wallace, 1997). However, there is little or no standardization between these webs, with the strength of links being expressed in many different ways. Berlow *et al.* (2004) list 11 definitions of "interaction strength" in use in the ecological literature. Inevitably, this lack of standardization mitigates against the detection of generalities, or the lack of them, when comparing among webs (McCann, 2000). This rather loose use of the term reflects, at least partially, the contingent nature of the field and the historical schism between the parallel schools of the ecosystem and community approaches. The important question for food web ecologists is: are these different definitions somehow related, such that they refer to similar phenomena? For instance, are population dynamics and energy flux linked? Interaction strengths defined or measured in different ways are not *necessarily* related, so comparisons between quantified webs constructed using the different approaches should be made with caution, unless some form of cross-validation can be carried out. For instance, while a single prey species might account for much of a predator's production, the population size of the prey may itself be unaffected. Conversely, a predator may consume a large proportion of the production of one prey species, while that prey contributes little to the overall production of the predator (e.g., if predators are subsidized by alternative food sources).

Producing quantified webs is extremely labor-intensive, and finding a single measure of interaction strength that is acceptable to both modellers and empiricists has yet to be achieved (Berlow *et al.*, 2004). For instance, it has been argued that energy flow webs cannot reveal whether consumers have negative effects on their food supply or competitors (e.g., Paine, 1988) and that field experiments are therefore required to unravel such causal relationships (e.g., Paine, 1992). However, conducting manipulations that include even a small fraction of the pairwise interactions within a web is unfeasible in all but the simplest systems. Even the relatively "simple" community of our study site (Broadstone Stream), which has a very restricted fauna due to its acidity, has over 700 links when both the macrofauna (animals retained on a 500 μm mesh) and meiofauna (animals that pass through a 500 μm mesh but are retained on a 42 μm mesh) are included (Schmid-Araya et al., 2002a). Further, to detect many indirect food web effects (e.g., trophic cascades) requires experiments that run for at least twice the generation time of the longest-lived organism within the system (Yodzis, 1988), which is clearly impracticable for most situations. Consequently, some authors have argued in favor of assessing linkage strength via detailed surveys rather than attempting to conduct unrealistic experiments (e.g., de Ruiter et al., 1995), and recent work has linked energy flux to interaction strength (Wootton, 1997). Although no truly complete community food web has been quantified to date, important progress has been made recently with semi quantitative measures, subsets of communities, and experimental systems (e.g., Paine, 1992; de Ruiter et al., 1995; Raffaelli and Hall, 1996; Benke et al., 2001; Emmerson and Raffaelli, 2004). Where attempts have been made at quantification, the shortcomings of connectance webs and the theoretical predictions derived from them have often been brought sharply into focus (McCann, 2000). The next logical—but logistically challenging—step is to produce detailed, fully-quantified, empirical food webs that are standardized sufficiently to allow the implicit connections between structural attributes and dynamics processes to be explored more fully, and for models to be parameterized from real data.

C. The Broadstone Stream Food Web

Broadstone Stream has had a long history of food web research (Hildrew et al., 1985; Lancaster and Robertson, 1995; Woodward and Hildrew, 2001), culminating in one of the most completely described food webs for any system (Schmid-Araya et al., 2002a). A major thrust of this research has been to increase taxonomic resolution and completeness by extending the lower size limit of the organisms included, to the meiofauna and beyond, and thence to test the effect on food web patterns. We collated all of the food web data collected from Broadstone to date to produce what we believe to be the most detailed connectance web yet described for any system, in order to examine body size constraints on web structure. However, with an emphasis on the soft-bodied meiofauna (e.g., rotifers, tardigrades, nematodes), overall replication, habitat coverage, and sampling extent are inevitably sacrificed in exchange for novel information on these small organisms. We are therefore restricted to examining static patterns in connectance webs because the dynamics of these small taxa are impossible to quantify empirically in the field.

Our primary objective in the present study was to quantify the Broadstone Stream food web. To do this, we made the strategic decision at the outset to exclude the "permanent meiofauna," or taxa that spend their entire life cycle within the meiofaunal size fraction of 42–500 μm (e.g., harpacticoids and rotifers), from the first attempt. Thus, we focused on quantifying the macrofaunal subweb, which—although omitting the permanent meiofauna—included representatives of the "temporary meiofauna" (i.e., macrofaunal taxa that spend only part of their entire life cycle within the meiofaunal size fraction, such as early instars of some of the smaller insect species). This focus enabled us to preserve samples (many soft-bodied permanent meiofauna can only be identified alive) and to substantially increase the extent and coverage of sampling. Therefore, the detection and assessment of energetically important links was improved, especially those of the larger, but less abundant, species near the top of the web.

We then used our data to construct a connectance web for each sampling occasion, thus addressing temporal variation in web structure. We were also able to construct yield-effort curves to examine seasonal and taxonomic influences on the estimation of food web statistics. Potentially important temporal patterns are often masked in other studies, because most webs are constructed from either a single sampling occasion or pooled over several occasions (but see Warren, 1989; Winemiller, 1990; Closs and Lake, 1994; Tavares-Cromar and Williams, 1996; Thompson and Townsend, 1999). Secondly, and most importantly, we sought to quantify the macroinvertebrate food web by calculating consumption rates of predators and relating their annual ingestion to the annual production of both predators and prey, thereby providing empirical estimates of the strength of feeding links. We then assessed the role of body size as a determinant of pattern (e.g., diet width) and process (e.g., ingestion rates, I/P) in the food web, in light of recent advances in theory that have implicated body size as a key driver in natural food webs (Warren, 1996; Williams and Martinez, 2000; Cohen *et al.*, 2003).

III. METHODS

A. Study Site

Broadstone Stream (51 °05′N, 0°03′E; 120 m above sea level) is a headwater of the River Medway in southeast England (see Hildrew and Townsend, 1976 for a detailed site description). The acidity of the stream (pH, 4.7–6.6) excludes fish and other vertebrates are extremely rare, resulting in an invertebrate-dominated food web (Woodward and Hildrew, 2001). The most detailed web described to date contains 128 species, including the permanent meiofauna (species always passing a mesh of 500 μm but retained on 42 μm), although there are only about 25 common macroinvertebrate species (Schmid-Araya *et al.*, 2002a; Woodward and Hildrew, 2002b). Among the common predators, there are three large species (*Cordulegaster boltonii*

Donovan, *Sialis fuliginosa* Pict. and *Plectrocnemia conspersa* [Curtis]) and three small species (the larvae of the tanypod midges *Macropelopia nebulosa* [Meigen], *Trissopelopia longimana* [Staeger] and *Zavrelimyia barbatipes* [Kieffer]). Detritivorous stoneflies and chironomids dominate the macrofaunal prey assemblage in winter and summer, respectively. The hyporheic zone (i.e., the interstitial habitat between the surface and groundwater, inhabited by the hyporheos) is very restricted, due to subsurface anoxic conditions, and rarely exceeds 5 cm depth (Rundle, 1988). Allochthonous detritus, in the form of coarse particulate organic matter (mostly woody debris and leaf fragments of > 1 mm diameter), is by far the most important basal resource (Dobson and Hildrew, 1992).

B. Estimation of Abundance and Biomass of Trophic Elements

Thirty randomly-dispersed Surber sample units (25×25 cm quadrant; mesh aperture $330\,\mu$m) were taken on each of six sampling occasions (May/June, August, October, December 1996, February and April 1997) to a depth of 5 cm. Samples were preserved immediately in 5% formalin. Because poor taxonomic resolution can confound comparisons among webs and grouping taxa is less meaningful in quantitative webs (Martinez, 1991; Hall and Raffaelli, 1993), we described all taxa to species where possible. The few species that could not be distinguished with certainty in benthic samples or predator guts were grouped to the next taxonomic level (usually genus). These groups were the oligochaete worms, nonpredatory tipulids, *Dicranota* sp., *Pedicia* sp., *Bezzia* sp., *Pisidium* sp., *Simulium* sp., *Helodidae* sp., and terrestrial invertebrates. Very rare taxa (i.e., <0.01% of mean annual standing stock) were excluded from the webs. Most species were univoltine and present only as larvae, so populations were not continually reproducing and recruitment was largely restricted to the summer. Generation times for the dominant taxa are given in Appendix 1. To estimate invertebrate biomass, linear body dimensions were measured and converted to dry mass using regression equations (listed in Woodward and Hildrew, 2002b).

The benthic density of coarse particulate organic matter (CPOM) was calculated as the oven-dried (60 °C) mass per sample unit, with leaves, woody debris, and fruiting bodies being weighed separately. Because length-weight regressions were not available for the terrestrial invertebrates, this basal resource was omitted from the biomass webs. However, most of the terrestrial prey found in predator guts consisted of oribatid litter mites or collembola which, because of their small size and rarity, probably accounted for relatively little energy flux within the web. Iron-bacteria (*Leptothrix* sp.) form ephemeral flocs, which carpet the bed during low flow, particularly in summer (Hildrew *et al.*, 1985). Diatoms are sparse in Broadstone Stream, and macroalgae are absent (Ledger, 1997); the heavily shaded channel, low pH, and presence of iron bacteria flocs prevent the formation of significant algal assemblage. The contributions of algae and iron bacteria to the web were not quantified, but were assumed to be small in comparison with detritus.

C. Construction of the Food Webs: Connectance Webs

We constructed a summary connectance web that included all links and species recorded over the six sampling occasions, excluding the permanent meiofauna. Connectance webs were also constructed for each sampling occasion, and are presented here as matrices (after Cohen et al., 1993a). Yield-effort curves were constructed for taxa and links on each sampling occasion. To assess potential underestimation of species and links, we fitted rectangular hyperbolae to these data: $y = B_{\max} . x/K + x$, where y is the number of taxa or links observed, x is the number of sample units processed (quadrats for species, guts for links), B_{max} is the asymptote for the number of taxa or links and K is the number of sample units needed to reach half B_{max}. Best-fit curves were generated using the GraphPad Prism Version 3.0 software package, and R^2 values were computed from the sum of the squares of the distances of the points from the best-fit curve determined by nonlinear regression (GraphPad Software Inc., 2000). Best-fit curves were derived iteratively, by varying the values of the variables to minimize the sum-of-squares.

We calculated several food web statistics for the connectance webs. Maximum chain length was the number of trophic elements (i.e., species or other taxonomic units) in the longest food chain from a basal resource to a top predator; where there were feeding loops, each cycle was counted once. Directed connectance was calculated as $C = L/S^2$ (Martinez, 1991), where L is the number of realized trophic links observed and S is the number of trophic elements in the web. We used this measure because it is less susceptible to variations in web size than other estimates of connectance (Martinez et al., 1999). Complexity was calculated as $SC = S(L/S \, [S - 1]/2))$ (after Polis, 1991). The mean number of links per species, d, was calculated as L/S. We excluded basal resources from calculations of all web statistics (except chain length) because detritus could not be classified to species. Suctorial predators were omitted from the calculation of web statistics because their feeding links could not be described (after Closs and Lake, 1994).

We also constructed a highly resolved summary food web (including the permanent meiofauna) that included every link and species recorded in Broadstone Stream since sampling began in the early 1970s (webs were collated from the current study; Hildrew et al., 1985; Lancaster and Robertson, 1995; Woodward and Hildrew, 2001; and Schmid-Araya et al.,

2002a). Because of the large number of samples used to construct this web, we assumed that asymptotes for the feeding links of the six most abundant predators had been reached (>1000 guts per species); these species were then used to examine relationships between predator body size and diet width. Basal resources were included in the calculation of food web statistics for this web to facilitate comparisons with earlier work.

D. Quantification of Feeding Links

The individual feeding links of the primary consumers (all detritivores with the exception of the very rare grazing mayfly, *Paraleptophlebia submarginata*) were not quantified. However, detritus is abundant throughout the year and not limiting as a food resource (Dobson and Hildrew, 1992). The predators used for gut contents analysis were *C. boltonii* ($n = 411$ guts), *S. fuliginosa* ($n = 450$ guts), *P. conspersa* ($n = 559$ guts), *M. nebulosa* ($n = 543$ guts), *T. longimana* ($n = 1039$ guts), *Z. barbatipes* ($n = 824$ guts) and three rarer species, the stonefly *Siphonoperla torrentium* (Pictet) ($n = 59$ guts) and the tipulids *Dicranota* sp. ($n = 102$ guts) and *Pedicia* sp. ($n = 24$ guts). Gut contents analysis was performed on individuals of all macroinvertebrate predators collected in the Surber samples on each occasion, except the two pentaneuriids (*T. longimana* and *Z. barbatipes*), which were randomly subsampled in August and October 1996 when they were extremely abundant. Subsampling reduced processing time but, nevertheless, over 400 pentaneuriid guts were analyzed in each of these two months.

The guts of the predators were dissected, mounted in euparal, and examined at $400\times$ magnification. Gut contents were identified from reference slides and the biomass of ingested prey was estimated from length-weight regressions from linear dimensions (listed in Woodward and Hildrew, 2002b). Because prey were generally consumed whole, or in large fragments, species could be identified relatively easily. Chironomid head capsule widths in the guts were reduced by 17%, to correct for flattening during mounting (after Hildrew and Townsend, 1982).

We wanted to use the gut contents data to estimate the *per capita* consumption rates (i.e., the number of prey per predator per day) in the food web links. Two problems were encountered: (1) the time during which consumed prey remain identifiable depends on temperature; and (2) the numbers observed in individual guts are highly variable and include many zeros. To deal with these problems, we used an approach based on that of Speirs *et al.* (2000). Let X be the mean number of prey per predator gut, and let τ be the characteristic residence time of an item of prey in the gut. If prey are ingested at rate I, then the rate of change of the numbers of recognizable prey is $dX/dt = I - X/\tau$. Thus, if food consumption and digestion are in balance, we have $dX/dt = 0$ and so

$$I = \frac{X}{\tau}$$

The first requirement for estimating I is therefore obtaining an appropriate measure of X, the mean number of prey per predator gut. Since the observed number of prey is frequently zero and highly variable, it makes sense to view it as a Poisson variable with mean X. If r_i is the observed number of prey in the ith gut analyzed, then this has likelihood (i.e., the probability of the observation assuming a mean X) of

$$l_i = \frac{X^{r_i}e^{-x}}{r_i!}$$

Thus, the likelihood ℓ of the whole data set of n guts is

$$\ell = \prod_{i=1}^{n} \frac{X^{r_i}e^{-x}}{r_i!}$$

which yields a negative log-likelihood L of

$$L = \sum_{i=1}^{n}[\ln(r_i!) - r_i \ln X + X]$$

Now, this has derivative with respect to X of

$$\frac{dL}{dX} = \sum_{i=1}^{n}\left(\frac{r_i}{X} - 1\right)$$

and this will be minimized when $dL/dX = 0$. Thus the maximum likelihood estimator of X is obtained by

$$X = \frac{\sum_{i=1}^{n} r_i}{n}$$

The second requirement for obtaining I is the gut residence time τ, which is known to be temperature dependent in a nonlinear fashion (Hildrew and Townsend, 1982). We assume the form

$$\tau = \tau_0 e^{-T/T_0}$$

where τ_0 and T_0 are constants, and T is the ambient temperature in °C during the period when the sample was taken. Thus, in calculating the ingestion rates for the seasonal webs, we used the mean stream temperature for the month under consideration. Annual *per capita* ingestion rates were then obtained by taking the mean of the seasonal values.

In order to determine τ_0 and T_0, we note that the Q_{10} for this process is, by definition, the ratio of the rates (*i.e.*, $1/\tau$) $10\,°C$ apart. Thus,

$$Q_{10} = \frac{\tau_0 e^{-T/T_0}}{\tau_0 e^{-(T+10)/T_0}}$$

and hence,

$$Q_{10} = e^{10/T_0}$$

Thus, if we know the Q_{10} and the gut residence time τ at a reference temperature, we can determine both T_0 and τ_0. In their study of fourth and fifth instar *Plectrocnemia conspersa*, Townsend and Hildrew (1977) report a Q_{10} of 2.3, which implies a value of $T_0 = 12\,°C$. They also found that the prey-recognition "half-life" at $13\,°C$ was 9.8 hours for stoneflies and 7.3 hours for chironomids. Since the half-life is $\tau \ln(2)$, we get values of τ_0 of 41.7 hours for stoneflies and 31.1 hours for chironimids. A few prey species were neither chironomids nor stoneflies, and for these we used the values for either the former or the latter depending on how similar they were in morphology. For example, *Sialis* larvae in the guts of predators were of similar size and degree of sclerotisation to stoneflies, while Ceratopogonidae (*Bezzia* sp.) were considered equivalent to chironomids. An additional problem was that recognition time in the gut is inversely related to predator biomass, and increases with prey biomass Hildrew and Townsend (1982). However, since the mean biomass of individual prey in a predator's gut also increases with predator biomass with a slope close to unity (Woodward and Hildrew, 2002b), we follow Speirs *et al.* (2000) in assuming that recognition time for ingested prey was constant across predator size classes.

The diet was characterized for all but two minor predator species: the larvae of *Platambus maculatus* (L.) (Dytiscidae) and *Bezzia* sp. (Ceratopogonidae) are suctorial predators, so the guts do not contain identifiable sclerotized material (cf. Closs and Lake, 1994). Because *P. maculatus* (species no. 7) was rare in Broadstone, it probably had little effect on prey populations. *Bezzia* sp. (species no. 11) was relatively abundant, but very small.

E. Construction of the Food Webs: Quantitative Webs

We constructed quantitative food webs, based on both density and biomass, on each of our six sampling occasions. Because reliable estimates of benthic density could not be obtained for very small individuals ($<10\,\mu g$) on each sampling occasion, they were excluded from the seasonal webs, although we were able to estimate the contributions of the temporary and permanent meiofauna to the annual production-ingestion web (see below). Feeding links in the seasonal webs were expressed as *per capita* consumption 24 h^{-1}, as a percentage of the numbers or biomass of each prey population (i.e., if the abundance of *P. conspersa* was 50 m^{-2} and, on average, each *C. boltonii* ate one individual 24 h^{-1}, this link would be assigned a value of 2%). Links to the basal resources were not quantified.

We further estimated both annual secondary production (of prey and predators) and annual ingestion (by predators) to produce a quantified measure of biomass flux through the web over the entire year. Stead *et al.* (2005) recently measured secondary production of the meiofauna and macrofauna in a nearby acid stream (Lone Oak) that contains a very similar species complement to Broadstone. Both streams also have comparable macroinvertebrate faunal densities: mean annual standing biomass is 0.66 g m^2 and 0.83 g m^2 in Lone Oak and Broadstone, respectively. Although the meiofauna accounted for 52% of total production in Lone Oak, most of this was due to small instars of macrofaunal species in the temporary meiofauna (38% total production); the permanent meiofauna, such as rotifers and tardigrades, contributed relatively little (14%). Because we could not measure production across all size classes of the temporary meiofauna directly in Broadstone due to logistic constraints (we used a 330 μm mesh; Stead *et al.* used a 42 μm mesh), we estimated the "missing" biomass (and numbers of individuals) in this web by assuming that the ratio of the mean annual biomass (and numbers) of individuals >10 μg: <10 μg was the same for identical macroinvertebrate taxa in the two streams. The data were split into these two body mass categories because all individuals of ≥10 μg were sufficiently large to be retained by the 330 μm mesh used in Broadstone. Similarly, the permanent meiofauna was assumed to account for the same proportion of total community production in both streams. We then multiplied the mean annual biomass of each taxon by its P/B ratio, as derived empirically in Lone Oak using the size-frequency method (Benke, 1993; Stead *et al.*, 2005), to estimate production in Broadstone.

Using P/B ratios measured in one system to predict production in another, comparable, system (e.g., Strayer and Likens, 1986) provides an alternative to the direct, but more labor-intensive, size-frequency method. We did employ the latter method, however, to directly measure the production of four of the larger Broadstone taxa for which we had reliable abundance data for all size classes. This enabled us to compare our calculated P/B ratios with those derived by Stead *et al.* (2005) in Lone Oak.

Gut contents data were used to estimate organic matter flux within the food web. Because some of the predators, particularly the smaller species, also fed on nonanimal prey, we estimated the contributions of these different food types to secondary production. We assumed that assimilation efficiency (AE, assimilation/ingestion) was 10% for coarse particulate organic matter

(CPOM, >1 mm diameter), 27% for fine particulate organic matter (FPOM, 50 μm – 1 mm diameter), 30% for algae, and 70% for animals (after Benke and Wallace, 1997). Annual ingestion of individual prey species by each predator was determined from both the predator's production and the percentage of the biomass in its diet represented by each prey species. Total ingestion by each predator was estimated, after Benke *et al.* (2001), as its production divided by the gross production efficiency (GPE), where GPE = AE × net production efficiency (NPE, production/assimilation). We assumed NPE to be 55%, after the studies of Smock and Roeding (1986) and Smith and Smock (1992); the latter study site was carried out in a headwater stream that contained similar predatory taxa to those in Broadstone, including *Cordulegaster* sp., and *Zavrelimyia* sp. GPE was estimated at 38.5% (i.e., 70 × 55%, after Benke *et al.*, 2001) for the exclusively predatory taxa (*C. boltonii, S. fuliginosa, P. conspersa, P. maculatus* and *Bezzia* sp.), within the 33–39% range reported for invertebrate predators by Slansky and Scriber (1982). The predatory taxa that supplemented their diets with nonanimal food had GPE values ranging from as low as 7.0% for *Pedicia* (which ingested large quantities of CPOM) to 30.8% for *M. nebulosa* (which ingested mostly animal prey but also FPOM and, to a lesser extent, algae). Production and ingestion rates were calculated per unit area of stream bed per year (g dry mass m^{-2} y^{-1}), and the quotient of annual ingestion/production was used to provide a quantitative estimate of "interaction strength."

IV. RESULTS

A. Summary Connectance Web (Including the Permanent Meiofauna)

This highly-resolved summary web, which includes all the food web data collected since the 1970s, contained 131 "species" and 842 links (Table 1). Addition of the permanent meiofauna to the equivalent macroinvertebrate-only summary web caused a slight decline in both complexity (13.11 to 12.93) and linkage density (6.45 to 6.42), a halving in connectance (0.104 to 0.049), but an increase in maximum chain length (12 to 15 species) (Table 1). This new summary web contained 17% more links than the web described by Schmid-Araya *et al.* (2002a), with 48 of the 121 additional links being from the invading top predator, *Cordulegaster boltonii*. Web size, however, increased by only 2.3%, following the inclusion of the mayfly *Paraleptophlebia submarginata* (Stephens), the amphipod *Niphargus aquilex* Schiödte and the isopod *Asellus meridianus* Racovitza, which were not recorded by Schmid-Araya *et al.* (2002a), probably because of the smaller sampling effort used in their study. Consequently, compared with Schmid-Araya *et al.*'s (2002a) highly resolved summary web, there was a

Table 1 Food web statistics for the Broadstone Stream food web at high and low (macrofauna only) resolution

	High resolution		Low resolution	
	Summary web	Schmid-Araya et al. (2002a)	Summary web (excludes permanent meiofauna)	Schmid-Araya et al. (2002a) (excludes permanent meiofauna)
No links (L)	842	721	400	319
Web size (S)	131	128	62	59
Directed connectance (C)	0.049	0.044	0.104	0.092
Links per species (d)	6.42	5.63	6.45	5.41
Complexity (SC')	12.93	11.35	13.11	11.00
Maximum chain length	15	12	12	9

The summary web represents data collated from all published sources in addition to the current study (see Methods); the summary web described by Schmid-Araya et al. (2002a) represents data collected over one year, 1996/97.

slight increase in connectance (0.044 to 0.049), but more marked increases in complexity (11.35 to 12.93) and linkage density (5.63 to 6.42).

The total number of links per predator was determined by simple logarithmic body-size relationships between predators and their prey (Fig. 1). For the six most intensively sampled predators, the log_{10} total number of predatory links decreased with log_{10} mean individual predator body size (μg) ($y = 1.78 - 0.049x$; $r^2 = 0.86$; $F = 24.52$; $p = 0.008$). The opposite was true, however, when the permanent meiofauna were excluded ($y = 1.48 + 0.035x$; $r^2 = 0.93$; $F = 55.85$; $p = 0.002$), suggesting the existence of both upper and lower size refugia for prey. Thus, more of the species in the web were vulnerable to the small-bodied tanypod Z. barbatipes than to the large-bodied dragonfly, C. boltonii (i.e., there is a lower size refugium from the dragonfly), whereas more macroinvertebrate prey are vulnerable to C. boltonii than to Z. barbatipes (i.e., there is an upper size refugium from the tanypod).

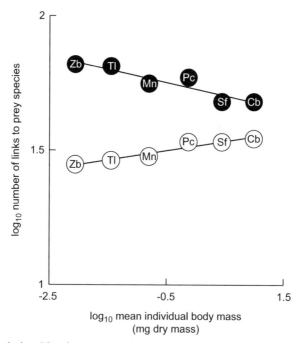

Figure 1   Relationships between predator body-mass and the number of feeding links to the macrofauna and permanent meiofauna within the highly-resolved summary food web for Broadstone Stream. The solid circles represent all links, including the permanent meiofauna; the open circles represent links to macrofaunal prey only. Predator codes: Cb, *Cordulegaster boltonii*; Sf, *Sialis fuliginosa*; Pc, *Plectrocnemia conspersa*; Mn, *Macropelopia nebulosa*; Tl, *Trissopeloia longimana*; Zb, *Zavrelimyia barbatipes*.

B. Seasonal Connectance Webs (Excluding the Permanent Meiofauna)

Although most members of the macroinvertebrate summary web (Fig. 2) were recorded on every sampling occasion, the number of links varied seasonally, being greatest in summer and declining progressively until the following spring (Table 2). Yield-effort curves suggested that sampling effort was sufficient to detect virtually every species, but not the total number of links, on each sampling date (Fig. 3). Rectangular hyperbolae described the curves for species and the links of the six dominant predators well, with high R^2 values (Table 3; mean $R^2 = 0.92$). For most predator species, the asymptote for the total number of links was not reached within individual sampling occasions, even when more than 200 guts were analyzed. Consequently, links from the rarest predators (*Pedicia* sp., *Dicranota* sp. and *Siphonoperla torrentium*) were those most severely underestimated. For the six most common predators, the asymptotic number of links increased with predator size (e.g., B_{max} equalled 27 and 9 for *C. boltonii* and *Z. barbatipes*, respectively), whereas the number of guts required to estimate 50% of B_{max}, called K, generally decreased with increasing predator size (trophic status) (e.g., K equalled 33 and 70 for *C. boltonii* and *Z. barbatipes*, respectively) (Table 3). Although predator size and abundance (sample size for guts) were inversely related and abundance varied seasonally (see also Woodward and Hildrew, 2002b), on average 63% of the predicted number of feeding links were recorded on each sampling occasion because of the compensatory effect of the inverse relationship between K and body size.

Web complexity, connectance, and links per species were greatest during summer, when abundance was highest and generations overlapped (Table 2). All these measures declined progressively until the following spring as the web became "simpler," and were always lower than in the summary web (e.g., each statistic in April was less than half the corresponding value in the summary web). Food chains included up to eight species (excluding loops) and were longest in summer and autumn. Mutual predation and cannibalism occurred among the dominant predators, especially during autumn, when the predator body size distribution was broadest. Omnivorous links were common: for example, in an eight-species food chain that linked *Cordulegaster boltonii* to the basal resources (terrestrial invertebrates), each species was preyed on directly by *C. boltonii*, including conspecifics. Further, the tanypods, particularly *Macropelopia nebulosa*, derived a portion of their diet from algae, especially during summer when this resource was most abundant, and FPOM was also frequently ingested, particularly by the smaller instars. Because FPOM was abundant throughout the year, the increased consumption of detritus in winter (e.g., 25% and 42% of *T. longimana* guts contained FPOM in August and February, respectively; $\chi^2 = 7.35$, $P < 0.01$) suggested

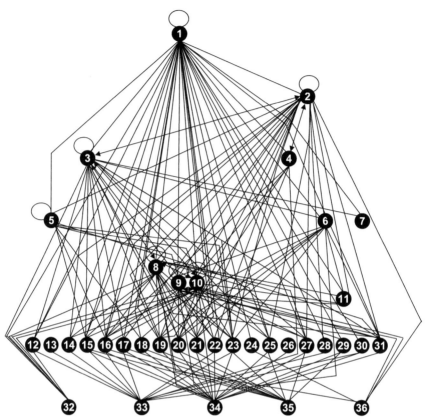

Figure 2 Summary connectance food web for the macrofaunal assemblage of Broadstone Stream (1996–1997). Double-headed arrows depict mutual predation, circular arrows cannibalism. Key to species: 1. *Cordulegaster boltonii* (Donovan); 2. *Sialis fuliginosa* Pict; 3. *Plectrocnemia conspersa* (Curtis); 4. *Pedicia* sp.; 5. *Siphonoperla torrentium* (Pictet); 6. *Dicranota* sp.; 7. *Platambus maculatus* (Pictet); 8. *Macropelopia nebulosa* (Meigen); 9. *Zavrelimyia barbatipes* (Kieffer); 10. *Trissopelopia longimana* (Staeger); 11. *Bezzia* sp.; 12. *Potamophylax cingulatus* (Stephens); 13. *Adicella reducta* (McLachlan); 14. Tipulidae (non-predatory); 15. *Nemurella pictetii* Klapalek; 16. *Leuctra nigra* (Olivier); 17. *Leuctra hippopus* Kempny; 18. *Corynoneura lobata* Edwards; 19. *Prodiamesa olivacea* (Meigen); 20. *Heterotrissocladius marcidus* (Walker); 21. *Micropsectra bidentata* (Goetghebuer); 22. *Brillia modesta* (Meigen); 23. *Polypedilum albicorne* grp.; 24. *Paraleptophlebia submarginata* (Stephens); 25. oligochaetes; 26. *Pisidium* sp.; 27. *Simulium* sp.; 28. Helodidae sp.; 29. *Niphargus aquilex* Schiödte; 30. *Asellus meridianus* Racovitza; 31. cyclopoids; 32. Terrestrial invertebrates; 33. CPOM; 34. FPOM; 35. Iron bacteria; 36. Algae.

Table 2 Food web matrices for Broadstone Stream (1996–1997)

	Summary web 1 2 3 4 5 6 8 9 10	May 1 2 3 4 5 6 8 9 10	August 1 2 3 4 5 6 8 9 10	October 1 2 3 5 6 8 9 10	December 1 2 3 4 5 6 8 9 10	February 1 2 3 4 5 6 8 9 10	April 1 2 3 4 5 6 8 9 10
1	1 0 0 0 0 0 0 0 0	0 0 0 0 0 0 0 0 0	1 0 0 0 0 0 0 0 0	1 0 0 0 0 0 0 0	1 0 0 0 0 0 0 0 0	1 0 0 0 0 0 0 0 0	1 0 0 0 0 0 0 0 0
2	1 1 1 1 0 0 1 0 0	0 1 1 0 0 0 0 0 0	1 1 1 1 0 0 0 0 0	1 1 0 0 0 1 0 0	1 1 0 0 0 0 0 0 0	1 1 0 0 0 0 0 0 0	1 0 0 0 0 0 0 0 0
3	1 1 1 0 0 0 1 0 1	1 1 1 0 0 0 0 0 0	1 1 0 0 0 0 0 0 1	1 1 1 0 0 0 0 0	1 1 1 0 0 0 0 0 0	1 1 1 0 0 0 0 0 0	1 0 1 0 0 0 0 0 0
4	1 1 0 0 0 0 0 0 0	1 0 0 0 0 0 0 0 0	0 0 0 0 0 0 0 0 0		1 1 0 0 0 0 0 0 0	1 0 0 0 0 0 0 0 0	1 0 0 0 0 0 0 0 0
5	1 1 0 0 1 0 0 0 0	1 0 0 0 0 0 0 0 0	1 0 0 0 0 0 0 0 0	1 0 0 0 0 0 0 0	1 1 0 0 0 0 0 0 0	0 0 0 0 0 0 0 0 0	1 0 0 0 0 0 0 0 0
6	1 1 1 0 0 0 0 0 0	0 0 0 0 0 0 0 0 0	0 1 0 0 0 0 0 0 0	0 0 1 0 0 0 0 0	1 0 0 0 0 0 0 0 0	1 0 0 0 0 0 0 0 0	1 0 0 0 0 0 0 0 0
8	1 1 1 0 0 0 1 0 1	1 0 0 0 0 0 0 0 0	1 1 1 0 0 0 0 0 0	1 1 1 0 0 1 0 1	0 1 0 0 0 0 1 0 1	0 0 0 0 0 0 1 0 0	1 0 0 0 0 0 0 0 1
9	1 1 1 0 1 1 1 1 1	1 1 1 0 0 1 1 1 1	1 1 1 0 0 0 1 0 1	1 1 0 0 1 0 1	1 1 0 0 0 1 0 1	1 0 1 0 1 1 1 1 1	1 1 0 0 0 0 0 0 0
10	1 1 1 1 1 0 1 1 1	1 1 0 0 1 0 1 0 1	1 1 1 1 0 0 1 0 1	1 1 1 1 0 1 1 1	1 0 0 0 1 0 1 1 1	1 0 0 0 0 0 1 0 1	1 0 1 0 1 0 1 0 1
12	1 1 1 0 0 0 1 0 0	1 0 1 0 0 0 0 0 0	1 0 1 0 0 0 0 0 0	1 1 1 0 0 0 0 0	0 1 1 0 0 0 0 0 0	1 1 1 0 0 0 0 0 0	1 0 1 0 0 0 0 0 0
13	0 0 0 0 0 0 0 0 0				0 0 0 0 0 0 0 0 0	0 0 0 0 0 0 0 0 0	
14	1 0 1 0 0 0 1 0 1	1 0 0 0 0 0 1 0 0	0 0 0 0 0 0 0 0 0	0 0 0 0 0 0 0 0	1 0 1 0 0 0 1 0 1	1 0 0 0 0 0 0 0 0	0 0 0 0 0 0 0 0 0
15	1 1 1 0 1 0 1 1 1	1 1 1 0 0 0 0 0 0	1 1 1 0 0 0 1 1 1	1 1 1 0 0 1 0 1	1 1 1 1 0 0 0 0 1	1 1 1 0 1 0 1 0 1	1 1 1 0 1 0 0 0 0
16	1 1 1 1 1 1 1 1 1	1 1 1 0 0 0 1 1 1	1 1 1 0 0 1 1 1 1	1 1 1 0 0 1 1 1	1 1 1 0 1 1 0 0 1	1 1 1 1 1 0 1 1 0	1 1 1 1 0 0 0 0 0
17	1 1 1 0 0 0 0 0 0	1 0 0 0 0 0 0 0 0	0 0 0 0 0 0 0 0 0	1 1 1 0 0 0 0 0	1 0 0 0 0 0 0 0 0	0 0 0 0 0 0 0 0 0	0 0 0 0 0 0 0 0 0
18	1 0 0 0 1 0 0 0 1		1 0 0 0 0 0 0 0 1	0 0 0 0 0 0 0 0	0 0 0 0 0 0 0 0 1	0 0 0 1 0 0 0 0 0	
19	1 1 1 1 0 1 1 1 1	0 0 1 0 0 0 0 0 0	1 1 0 0 0 0 1 0 1	1 1 0 0 1 0 0	0 1 0 0 0 0 1 1 1	1 1 0 1 0 1 1 0 0	1 1 1 0 0 1 0 1 1
20	1 1 1 0 0 1 1 1 1	1 0 1 0 0 0 1 1 1	1 1 1 0 0 1 1 1 1	1 1 1 0 0 1 1 1	0 1 1 0 0 0 1 1 1	1 1 1 0 0 1 1 0 0	1 1 1 0 0 1 0 1 1
21	1 1 1 0 0 1 1 1 1	0 1 0 0 0 0 1 1 1	1 1 1 0 0 1 1 1 1	1 1 1 0 0 1 0 1	0 0 1 0 0 0 0 1 1	1 1 0 0 0 1 0 0 1	0 0 0 0 0 0 1 0 1
22	1 1 1 1 0 0 1 1 1	1 0 0 0 0 0 1 1 1	1 1 0 1 0 0 1 1 1	1 1 1 0 0 0 0 1	1 1 0 0 0 0 1 0 0	1 0 0 0 0 0 0 0 0	1 0 0 0 0 0 0 0 0
23	1 1 1 0 1 1 1 1 1	1 1 1 0 1 1 1 1 1	1 1 0 0 0 0 1 1 1	1 0 0 0 0 1 0 1	1 0 0 1 0 0 0 0 1	0 1 1 0 0 1 1 1 1	1 1 1 1 1 0 0 1 1
24	1 0 0 0 0 0 0 0 0	0 0 0 0 0 0 0 0 0		1 0 0 0 0 0 0 0	0 0 0 0 0 0 0 0 0	0 0 0 0 0 0 0 0 0	
25	1 1 1 0 0 0 1 0 1	0 0 0 0 0 0 0 0 0	1 1 0 0 0 0 1 0 1	1 0 1 0 0 0 0 0	0 0 0 0 0 0 0 0 1	0 0 0 0 0 0 0 0 0	0 0 0 0 0 0 0 0 0
26	0 0 0 0 0 0 0 0 0	0 0 0 0 0 0 0 0 0	0 0 0 0 0 0 0 0 0	0 0 0 0 0 0 0 0	0 0 0 0 0 0 0 0 0	0 0 0 0 0 0 0 0 0	0 0 0 0 0 0 0 0 0
27	1 1 1 0 0 0 1 1 1	0 0 1 0 0 0 0 0 0	1 1 1 0 0 0 1 0 1	0 0 0 0 0 0 0 0	1 0 1 0 0 0 0 1 1	1 0 0 0 0 0 1 0 0	0 0 0 0 0 0 0 0 0
28	1 1 0 0 0 0 0 0 0	0 0 0 0 0 0 0 0 0	0 0 0 0 0 0 0 0 0		1 1 0 0 0 0 0 0 0	0 0 0 0 0 0 0 0 0	0 0 0 0 0 0 0 0 0
29	1 0 0 0 0 0 0 0 1	0 0 0 0 0 0 0 0 0	0 0 0 0 0 0 0 0 1		0 0 0 0 0 0 0 0 0		0 0 0 0 0 0 0 0 0
30	1 0 0 0 0 0 0 0 0	0 0 0 0 0 0 0 0 0			0 0 0 0 0 0 0 0 0		0 0 0 0 0 0 0 0 0
31	1 1 1 0 1 1 1 1 1	0 0 1 0 0 0 1 1 1	1 1 1 0 0 0 1 1 1	1 1 1 0 0 1 1 1	1 1 1 0 1 0 1 1 1	1 0 1 0 0 0 0 0 1	1 1 1 0 0 0 1 0 1
Links per species (*d*)	4.52	2.33	3.35	3.00	2.59	2.33	1.88
Directed connectance	0.16	0.09	0.13	0.13	0.09	0.09	0.07
Complexity (*SC*)	9.36	4.84	6.96	6.26	5.36	4.85	3.92
Max chain length	8	7	8	8	8	7	6

Columns represent predators, rows represent prey. 1/0 represents presence/absence of a feeding link.

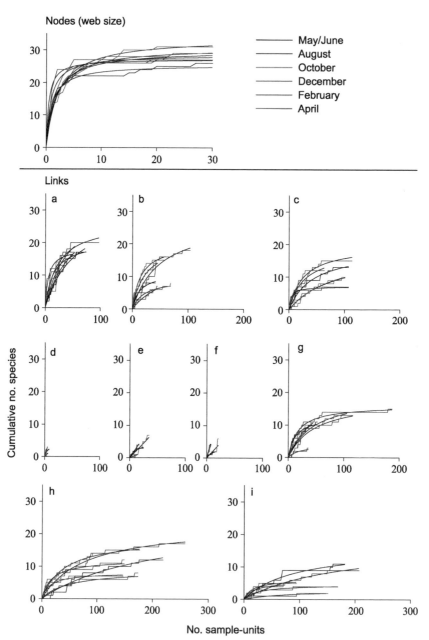

Figure 3   Yield-effort curves for the cumulative number of species included in the web as nodes (top panel) and as feeding links from prey to predators recorded within Broadstone Stream during 1996–97. Predators: a, *C. boltonii*; b, *S. fuliginosa*; c, *P. conspersa*; d, *Pedicia* sp.; e, *Dicranota* sp.; f, *S. torrentium*; g, *M. nebulosa*; h, *T. longimana*; i, *Z.*

Table 3 Predicted asymptotes (B_{max}) for the number of prey species consumed per predator species, and the number (K) of samples (benthic quadrats for species; guts for links) required to detect 50% of the predicted asymptote for the total number of feeding links

Species[b]	C. boltonii	S. fuliginosa	P. conspersa	M. nebulosa	T. longimana	Z. barbatipes
B_{max}						
May	25.53	31.98	11.19	18.61	12.73	24.86	16.37
Aug	27.20	26.58	27.87	18.28	16.08	22.94	17.91
Oct	28.54	34.55[a]	21.34	19.75	18.08	12.11	4.68
Dec	33.75[a]	33.53[a]	20.4	18.08	23.61	18.88	5.64
Feb	30.89	18.34	9.32	18.76	19.55	8.14	6.30
Apr	29.9	20.6	16.41	7.07	3.72	8.95	3.14
Mean	**29.30**	**27.60**	**17.75**	**16.76**	**15.63**	**15.98**	**9.01**
K						
May	1.01	55.2	13.68	42.39	12.25	209.6	92.28
Aug	0.45	23.15	50.37	27.04	16.81	78.65	176.3
Oct	0.85	44.57	23.66	25.7	47.09	18.25	25.79
Dec	2.26	52.3	15.88	83.95	53.49	41.17	17.55
Feb	1.81	5.909	21.97	88.3	46.25	15.2	16.08
Apr	1.64	15.09	73.9	3.73	17.6	51.11	89.38
Mean	**1.34**	**32.70**	**33.24**	**45.19**	**32.25**	**69.00**	**69.56**
Mean R^2	**0.98**	**0.94**	**0.94**	**0.94**	**0.92**	**0.95**	**0.84**

[a]B_{max} values for links that exceed the total number of invertebrate taxa included in the summary web (very rare species were excluded from the web; see Methods).
[b]Species ranked from left to right in order of decreasing body size.

that in summer, when small prey were most abundant, the tanypods became actively more predatory and less detritivorous.

There were clear trivariate relationships between body size, abundance, and web structure (after Cohen *et al.*, 2003), with most of the consumption flowing from smaller, more abundant prey to larger, rarer predators (i.e., energy moves upwards and to the left in Fig. 4a). Averaging across all the links in the web, predators were about one order of magnitude larger, by mass, than their prey (8–13 times larger, when cyclopoids were excluded or included in the web, respectively). These "rules" were broken in a few instances where predators fed on prey that were (on average) larger than themselves (Fig. 4b), but these links were rare and reflected seasonal and ontogenetic shifts in the relative body sizes of predators and prey, as described below.

barbatipes. Sample-unit (x-axis) for feeding links = 1 gut; for nodes = 1 Surber sample (25 cm × 25 cm quadrant). The black curves are models fitted to the data for each month and are rectangular hyperbolae (see text and Table 2).

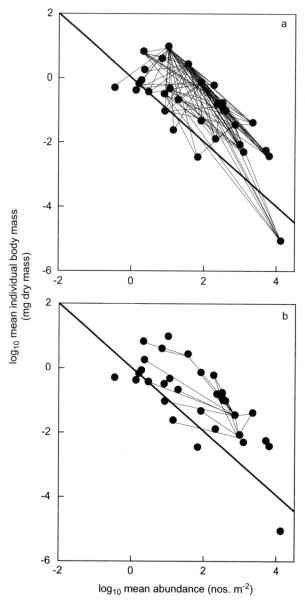

Figure 4 Summary food web for Broadstone stream, plotted on axes of mean abundance (x-axis) and mean body-mass (y-axis), after Cohen *et al.* (2003). Links that represent consumption where the predator is larger than its prey are shown in the top panel (a), whereas links that represent the opposite are shown in the lower panel (b). The dashed line, with a slope of -1, represents a biomass ratio that is equivalent between a consumer and resource: links with slopes that are more negative than -1 represent instances where a consumer has a greater biomass abundance than its prey (assuming the consumer is above and to the left of its prey in this plane, as in the top panel).

C. Quantitative Webs

There were strong seasonal shifts in invertebrate abundance. Peak recruitment after summer oviposition was followed by a progressive decline until spring for most taxa (Fig. 5). Among the basal resources, terrestrial detritus was extremely abundant, with little difference among months beyond a slight peak in February (which also corresponded with a slight winter peak in invertebrate abundance) (Fig. 5). Terrestrial invertebrates were most abundant during autumn, presumably due to inputs via leaf-fall, but were always rare relative to the aquatic fauna. Iron bacteria were present throughout the year, but only formed dense flocs in the summer. Algae were not visible in the stream to the naked eye, but ingestion of diatoms by the tanypods revealed that a limited algal biofilm persisted throughout the year. These seasonal changes in the availability of consumers and resources were reflected in the marked temporal changes in the magnitude and distribution of ingestion rates of prey by predators. Large numbers, especially of the smaller temporary meiofauna, were consumed in the summer, both *per capita* and per unit area (Table 4), although most links accounted for a relatively small proportion of total consumption (Fig. 6).

The presence/absence and strength of individual links varied seasonally as prey species entered or exited the different size ranges that could be handled by the respective potential predators (Figs. 7 and 8). For example, the *Macropelopia nebulosa* population ate equivalent to 14% (0.05% *per capita*) of the *Nemurella pictetii* population per 24 hours in August, but this link was not detected in April, when these prey were mostly too large to be handled by the predator (Fig. 7). Further, because of the negative relationship between mean abundance and body size (Fig. 4), often the numerical webs were effectively mirror-images of the biomass webs. The taxa and links that dominated the former, in terms of benthic abundance and ingestion rates, were usually relatively insignificant in the biomass webs, and *vice-versa*. The distribution of biomass among species within the webs, however, was less variable over time than in the numerical webs (Figs. 7 and 8), as individual growth mitigated the post-oviposition decline in numbers (Fig. 4). Within the predator guild, the large species dominated standing biomass throughout the year, whereas the smaller tanypods dominated numerically (Figs. 7 and 8). The relative importance of small and large predators varied seasonally, however, with the tanypods contributing considerably less to numbers or biomass per unit area other than in summer and autumn. Similar shifts occurred within the prey assemblage, with small species (chironomids) dominating in summer and large species (stoneflies) in winter (Figs. 7 and 8).

Although the tanypods ate a similar number of prey per unit area to the larger predators during summer and autumn (Table 4), they consumed considerably less biomass throughout the year because, being small

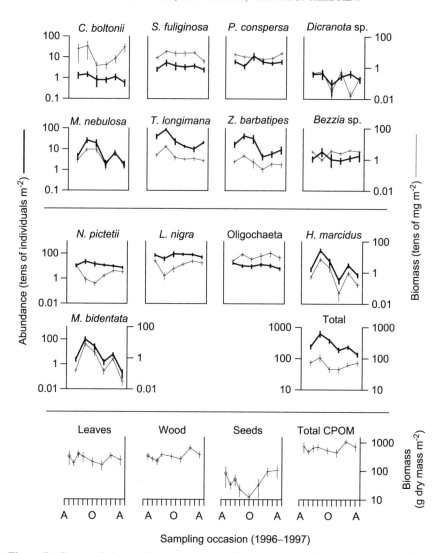

Figure 5   Seasonal changes in the mean (±1SE) abundance and biomass of the major macrofaunal consumers and basal resources in Broadstone Stream (1996–1997). Top panels represent predators, middle panels primary consumers, and lower panels represent the dominant basal resource, detritus (open circles represent additional sampling dates that were not used to estimate invertebrate abundance).

themselves, they ate much smaller prey (Figs. 7 and 8). *Per capita* ingestion, ingestion/m², the number of feeding links, and web complexity all fell markedly between August and April (Tables 1 and 4), tracking the decline in invertebrate abundance and shifts in the size spectrum. However,

Table 4 Estimated daily ingestion of prey in Broadstone Stream, 1996–1997

Month	Predator	Ingestion per predator (nos. eaten *per capita* 24 h^{-1})	Ingestion per unit area (nos. eaten m^{-2} 24 h^{-1})
May	Cb	0.73	9.75
	Sf	0.80	20.94
	Pc	1.33	81.59
	Mn	0.86	37.90
	Tl	0.19	76.96
	Zb	0.45	76.89
	Total	**5.09**	**312.33**
Aug	Cb	9.52	141.71
	Sf	4.22	227.43
	Pc	1.09	40.23
	Mn	4.31	1136.11
	Tl	1.83	806.25
	Zb	0.61	262.94
	Total	**25.98**	**2656.70**
Oct	Cb	1.88	15.01
	Sf	1.06	40.74
	Pc	0.68	77.07
	Mn	1.20	234.80
	Tl	1.75	268.10
	Zb	0.60	58.58
	Total	**8.44**	**706.57**
Dec	Cb	0.67	5.39
	Sf	1.00	33.44
	Pc	0.20	12.14
	Mn	0.73	14.29
	Tl	0.71	66.07
	Zb	0.19	8.36
	Total	**4.24**	**146.16**
Feb	Cb	0.87	9.77
	Sf	0.67	25.29
	Pc	0.28	14.15
	Mn	0.40	26.00
	Tl	0.37	37.91
	Zb	0.16	9.83
	Total	**5.03**	**140.32**
Apr	Cb	2.46	14.59
	Sf	0.83	20.75
	Pc	0.62	36.84
	Mn	0.17	3.15
	Tl	0.36	36.22
	Zb	0.02	2.17
	Total	**5.76**	**119.09**

The six dominant predators are shown separately; monthly totals include all predators. Predator codes: Cb, *Cordulegaster boltonii*; Sf, *Sialis fuliginosa*; Pc, *Plectrocnemia conspersa*; Mn, *Macropelopia nebulosa*; Tl, *Trissopeloia longimana*; Zb, *Zavrelimyia barbatipes*.

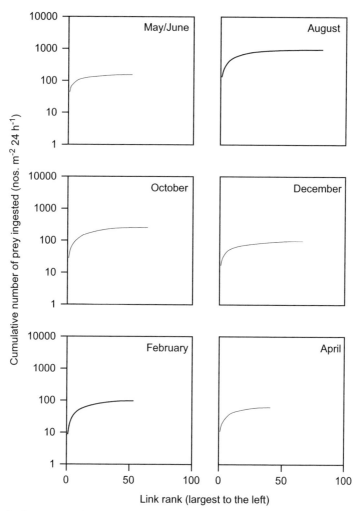

Figure 6 Structure of the Broadstone Stream food web: Feeding links. Links are ranked along the x-axis in order of decreasing magnitude (nos. individuals consumed $m^{-2}\,24\,h^{-1}$).

although overall consumption declined progressively from summer until spring, when expressed as a percentage of the total numerical standing stock (equivalent to 16.1% of the total macrofaunal standing stock ingested $m^{-2}\,24\,h^{-1}$ day in August, falling to 5.5% in April), the remaining links often increased in magnitude, so that the degree of skew within the web increased as complexity declined. Ingestion rates were skewed both among and within predator species and did not necessarily reflect prey density (i.e., the

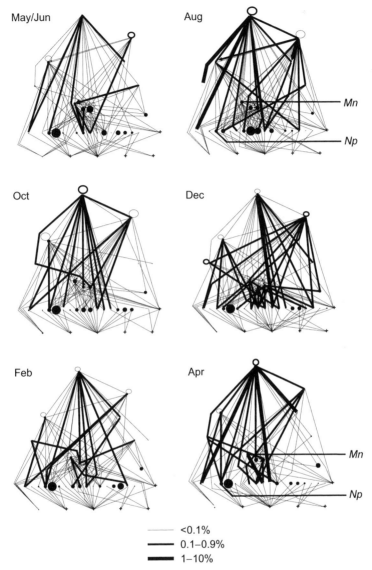

Figure 7 Quantified food webs representing numbers of macrofaunal prey (individuals $> 10\,\mu g$) eaten *per capita* $24\ \mathrm{h}^{-1}$ (as a percent of numbers m^{-2}) during 1996–1997. The area of each circle is proportional to total numerical standing stock within sampling occasions (see Fig. 5 for absolute values). Links to basal resources and the meiofaunal cyclopoids were not quantified (see Fig. 2 for comparison with connectance web and identity of taxa).

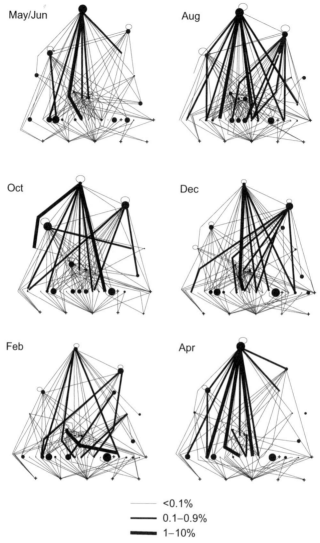

Figure 8 Quantified food webs representing biomass of macrofaunal prey (individuals > 10 μg) eaten *per capita* 24 h⁻¹ (as a percent of biomass m⁻²) during 1996–1997. The area of each circle is proportional to total standing biomass within sampling occasions (see Fig. 5 for absolute values). Basal resources are denoted nominally by '+'. Links to basal resources and the meiofaunal cyclopoids were not quantified (see Fig. 2 for comparison with connectance web and identity of taxa).

predators fed "selectively" on different portions of the size spectrum). In addition, there was a secondary effect of encounter rate increasing the predators' relative consumption of prey. For example, *per capita* consumption of the highly mobile, epibenthic stonefly *N. pictetii* was far higher than for the slow-moving, interstitial species *L. nigra*, despite the numerical dominance of the latter within the web (Fig. 7).

D. Annual Production and Ingestion Web

Total annual production for the entire community, excluding the permanent meiofauna, was estimated at 4.58 g m^{-2} y^{-1}. Of this total, 33% (1.52 g m^{-2} y^{-1}) was accounted for by the predator guild, although it should be kept in mind that many of these were omnivorous and often consumed the basal resources directly (Table 5). The three tanypod species accounted for equivalent to half the production (total, 0.42 g m^{-2} y^{-1}) of the three dominant large species (total, 0.80 g m^{-2} y^{-1}), despite being 15 times more abundant numerically. Production of the permanent meiofauna was estimated to account for an additional 14% (after Stead *et al.*, 2005), thus our estimate of total secondary production was 5.22 g m^{-2} y^{-1} (i.e., $4.58 + 0.64$ g m^{-2} y^{-1}). There was marked variation among species (in terms of their relative contributions to production) with the larger species, especially the predators and detritivorous stoneflies, being relatively productive per unit area despite having generally low P/B ratios (Table 5). Note that our estimated values of P/B for three large species were similar to those for the same taxa in Lone Oak calculated by Stead *et al.* (2005). Conversely, the detritivorous chironomids, with some of the highest P/B values (e.g., 10.12 for *Heterotrissocladius marcidus*), accounted for only 22% of total annual macrofaunal production despite their numerical dominance (41% and 70% of total and macrofaunal abundance, respectively), because of their low biomass per unit area. Similarly, the cyclopoids, with high P/B ratios (11.85) accounted for only 0.03% of production but 41% of total benthic density. The only species conventionally accepted as a "grazer," the mayfly *P. submarginata*, contributed a negligible 0.08% to prey production, suggesting the dominance of detritus as the principal basal resource: detritivores accounted for >99% of primary consumer production. *Pisidium* sp. and *Simulium*, the only filter-feeding taxa, accounted for 5.6 and 0.6% of prey production, respectively, but the former were not eaten by any of the predators. Consequently, the vast majority of energy reached the higher trophic levels via detrital food chains and by the processing of CPOM by shredders (mostly stoneflies) or FPOM by deposit feeders (mostly chironomids). Detritus was unlikely to be a limiting resource at any time throughout the year; assuming an assimilation efficiency of 10% (after Benke and Wallace,

Table 5 Mean annual production (dry mass), biomass, abundance and P/B ratio estimates for Broadstone Stream during 1996–1997[b]

Species	Production $(g\ m^{-2}\ y^{-1})$	Biomass $(g\ m^{-2})$	Abundance $(nos\ m^{-2})$	Body mass (mg)	P/B
Oligochaeta	0.782	0.112515	185	0.60807	6.95
Leuctra nigra	0.616	0.091976	2,208	0.04166	6.70
Prodiamesa olivacea	0.396	0.055252	320	0.17275	7.16
Sialis fuliginosa[a]	0.370	0.098786	36	2.75631	3.75 (3.72)
Heterotrissocladius marcidus	0.295	0.029145	5,104	0.00571	10.12
Plectrocnemia conspersa[a]	0.261	0.060629	82	0.73961	4.30 (5.80)
Micropsectra bidentata	0.241	0.023912	6,293	0.00380	10.07
Nemurella pictetii	0.239	0.034748	341	0.10188	6.87
Bezzia sp.	0.205	0.038367	393	0.09760	5.35
Macropelopia nebulosa	0.194	0.035855	228	0.15711	5.42
Cordulegaster boltonii[a]	0.173	0.099326	10	9.72509	1.74 (absent)
Pisidium sp.	0.172	0.040518	311	0.13034	4.24
Trissopelopia longimana	0.161	0.025646	722	0.03553	6.27
Potamophylax cingulatus[a]	0.137	0.027995	7	4.08489	4.89 (4.02)
Zavrelimyia barbatipes	0.060	0.008414	974	0.00863	7.14
Polypedilum albicorne	0.058	0.006317	1,242	0.00509	9.14
Pedicia sp.	0.049	0.014521	2	6.80678	3.38
Dicranota sp.	0.036	0.005356	11	0.47485	6.75
Brillia modesta	0.028	0.002718	208	0.01306	10.20
Leuctra hippopus	0.028	0.004135	19	0.21504	6.70
Simulium sp.	0.019	0.0039	82	0.04741	4.94
Helodidae	0.017	0.004155	2	1.82649	4.12
Siphonoperla torrentium	0.012	0.002584	8	0.32587	4.51
Niphargus aquilex	0.009	0.001114	3	0.38013	8.12
Asellus meridianus	0.008	0.001625	2	0.86680	5.00
Tipulidae	0.005	0.000784	8	0.09493	6.75
Platambus maculates	0.004	0.0011	2	0.68738	3.97
Diptera spp.	0.004	0.000563	1	0.42209	6.75
Paraleptophlebia submarginata	0.002	0.000344	14	0.02433	6.92
Corynoneura lobata	0.002	0.000238	67	0.00354	7.09
Cyclopoids	0.001	0.000117	13,196	0.00001	11.85
Adicella reducta	0.001	0.000176	0.3	0.50769	4.02
Sum	**4.58**	**0.832**	**32,082**		

[a]larger taxa without temporary meiofaunal stages for which P/B ratios were measured directly in Broadstone (values in parentheses are estimates for Lone Oak; see Methods). Taxa are ranked in order of descending production.
[b]Permanent meiofauna excluded except for cyclopoids.

1997), the entire shredder guild would have consumed equivalent to only 1.5% of the mean annual standing stock of CPOM.

The summary quantified food web illustrates the skewed contributions to both annual production (species nodes) and ingestion (feeding links) within the web (Fig. 9): a few taxa accounted for the majority of secondary production and most links were weak (I/P < 0.01). The composition of predator diets was determined by the relative sizes of predators and prey. The three large

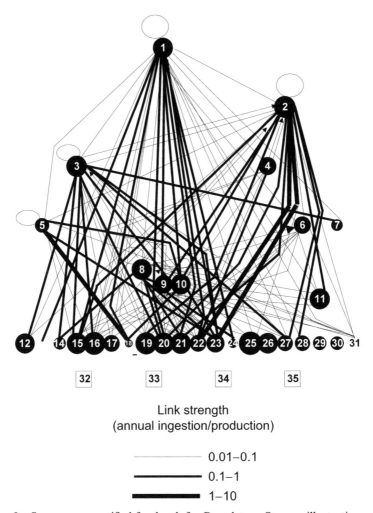

Link strength
(annual ingestion/production)

———————— 0.01–0.1

———— 0.1–1

■■■■■ 1–10

Figure 9 Summary quantified food web for Broadstone Stream, illustrating annual secondary production of predators and prey (proportional to circle area) with the strength of feeding links expressed as annual ingestion/annual production (I/P). Links to basal resources were not quantified.

predators preyed mostly on large prey (stoneflies, caddis, and other large predators: 40–65% of annual ingested prey biomass), whereas the smaller
predators took mostly small prey (chironomids: 70–85% of annual ingested
prey biomass). Although the overall distribution of the strength of individual feeding links (i.e., log_{10} I/P) within the entire web was positively skewed,
the position of individual predator species within this distribution reflected
differences in body mass (trophic status) (Fig. 10): the smaller predators had
fewer and weaker links to macrofaunal prey than the larger species, despite
their numerical dominance. However, recall that they had many more links
to the permanent meiofauna than did the larger predators (Fig. 1).

There were significant log-log correlations (all at $r > 0.5$ and $P < 0.005$)
between mean annual abundance, mean and maximum body mass, total
annual ingestion/production (I/P), P/B, and the frequency of ingestion (nos
ingested m^{-2} and *per capita*), suggesting that the trivariate relationships seen
in Figure 4 could be extended further to reveal multivariate relationships in
quantified webs (Table 6). The strength of predation pressure (as I/P) exerted
on prey species was negatively correlated with log_{10} mean (and maximum)
prey body mass and positively correlated with prey abundance. We calculated
the total consumption of each prey species by summing total ingestion across
all of its predators, to provide a single measure of "susceptibility" to predators for each species in the web (expressed as log_{10} total I/P). Most of the
production of the smaller taxa was consumed by the predators (Fig. 11).
Thus, larger prey species had relatively strong "bottom-up" effects on predator production, but suffered weaker "top-down" effects, in terms of the
proportion of their annual production that was eaten, when compared with
smaller species. For some of the very small taxa, ingestion exceeded production, possibly due to sampling errors, an overestimation of mean water
temperature, and/or external subsidies (e.g., drift). In terms of relating the
strength of links (total I/P) to their frequency of occurrence (numbers
ingested *per capita* $24 h^{-1}$), there was a positive log-log correlation ($r =
0.65$; $p < 0.001$), as stronger links were those most frequently observed.
This therefore suggested a relationship between the connectance and
quantified webs, in that increasing sampling effort simply leads to a greater
proportion of weak links being included in a web.

V. DISCUSSION

A. Connectance Webs

Schmid-Araya *et al.* (2002b) have recently documented a decline in connectance with increasing web size in a suite of well-characterized stream food
webs. While we cannot rule out a possible role of sampling artefact, our data

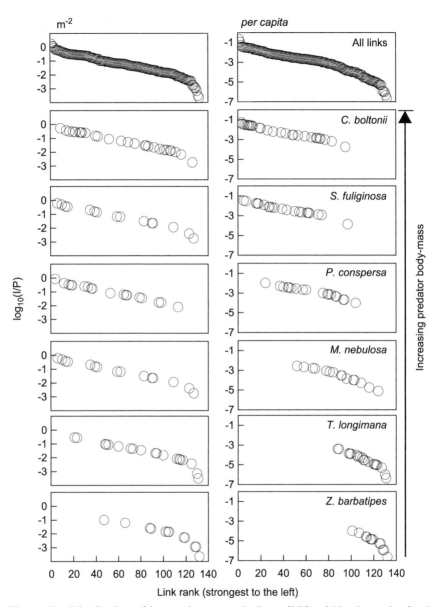

Figure 10 Distribution of interaction strength ($\log_{10}$ [I/P]) within the entire food web (top two panels) expressed as m^{-2} (left) and *per capita* (right). Each point represents an individual link within the food web. The positions of the individual links from the six dominant predator species within the overall distribution are shown in the 12 lower panels (m^{-2}, left panels; *per capita*, right panels).

Table 6 Pearson Product Moment Correlations (r) for $\log_{10}$-$\log_{10}$ relationships among biological and ecological traits and measures of interaction strength for the macroinvertebrate taxa in the quantified Broadstone Stream food web (Table 1)

	Nos m^{-2}	Mean body mass	Max body mass	P/B	I/P	I/P per capita	Nos. eaten m^{-2} 24 h^{-1}
Mean body mass	−0.74						
	−0.77						
Max body mass	−0.65	0.88					
	−0.69	0.88					
P/B	0.58	−0.80	−0.78				
	0.69	−0.86	−0.84				
I/P	0.59	−0.81	−0.77	0.59			
	0.67	−0.88	−0.83	0.81			
I/P per capita	0.52	−0.66	0.67	*0.37	0.89		
	0.60	−0.81	−0.76	0.69	0.90		
Nos. eaten m^{-2} 24 h^{-1}	0.94	−0.68	−0.61	0.67	0.61	**0.48	
	0.95	−0.68	−0.61	0.70	0.68	0.58	
Nos. eaten per capita 24 h^{-1}	0.92	−0.59	−0.57	0.50	0.65	0.63	0.93
	0.93	−0.59	−0.58	0.59	0.64	0.61	0.95

Pisidium sp, which were not eaten by any predator species, were excluded. The italicized values show correlations where the two predominantly hyporheic taxa, oligochaetes and *N. aquilex*, were excluded. All correlations were significant at $P < 0.005$, except those denoted by *($P < 0.05$) or **($P < 0.01$).

provide clear evidence for a mechanism that could account for the pattern obtained. Body size constraints had strong effects on predator diet width and this created a degree of compartmentalization between the permanent meio-faunal and macrofaunal subwebs in Broadstone Stream. Because the larger predators could not perceive or handle very small prey and, conversely, very large prey were invulnerable to small predators, size-refugia existed at both extremes of the size spectrum. Thus, speciose food webs including very small species are inevitably less richly connected than webs containing macrofauna alone, as was found here (Table 3). Essentially, there are no direct links between species at the two extremes of the food web, as was proposed in the "size disparity" hypothesis of Hildrew (1992) and Schmid-Araya *et al.* (2002b). The permanent meiofauna are rarely included in freshwater food webs (but see Schmid-Araya *et al.*, 2002a,b), and yet may provide an important energy source for predatory invertebrates, particularly in their early life stages (Woodward and Hildrew, 2002a). The highly-resolved summary web for Broadstone Stream contained 261 predatory links between the macrofauna and the permanent meiofauna. Ultimately, although the

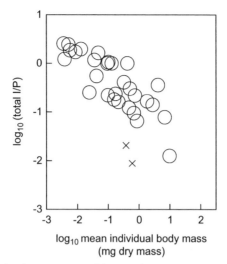

Figure 11   Prey body-size versus total interaction strength (total $\log_{10}$ [I/P]). Each datapoint represents the sum of annual ingestion/predation for each species (i.e., the total consumption of that species by all predators within the web). Two taxa that are predominantly hyporheic, *Niphargus aquilex* and oligochaete worms, and which were excluded from the second set of correlations in Table 6 (values in italics), are denoted by crosses, rather than circles.

task is daunting, by incorporating the permanent meiofauna into quantified webs, we should gain a better understanding of the trophic importance of these little-known taxa in freshwaters.

Our results support the findings of the few other studies that have examined temporal variations in web structure, which have also demonstrated striking seasonality (e.g., Warren, 1989; Winemiller, 1990; Closs and Lake, 1994; Tavares-Cromar and Williams, 1996; Thompson and Townsend, 1999). This seasonality is, unfortunately, inevitably masked in the summary webs that dominate much of the food web literature (e.g., catalogues in Cohen, 1978; Pimm, 1982; and Williams and Martinez, 2000). Web complexity in Broadstone was greatest in summer, when the availability of specimens for gut contents analysis was highest, and this might suggest that some of the seasonal patterns in web structure reported here (and elsewhere) were simply artefacts of sampling effort. However, there were no clear seasonal biases in the degree of our underestimations of food web links: about 60% of the predicted total number of links for the six dominant predators were detected on each occasion. This lack of any obvious seasonal methodological bias might be because, unlike many other studies, our sampling effort was relatively high and scaled to benthic abundance, rather than a fixed number of guts (Tavares-Cromar and Williams, 1996; Townsend et al., 1998; Thompson and Townsend, 1999). The absence of taxonomic bias in the detection of links (at least among the dominant predators) reflected a compensatory effect between predator body size and food web links in that, although the larger predators were rarer than their smaller counterparts, fewer needed to be sampled to characterize the diet. The *total* number of links detected on each sampling occasion, however, would clearly have been higher if gut contents analysis had been more exhaustive, and was underestimated even in the summary web because of the rarity of some predators (i.e., *Pedicia, Dicranota* and *S. torrentium*). However, the projected number of samples required to describe every individual feeding link within the web on a single sampling occasion would impose excessive disturbance upon the system and, because of the positive correlation between the frequency of occurrence and linkage strength (as I/ P), many of the very rare links would probably represent only very weak interactions (at least as measured by energy flow). This cannot always simply be assumed to be true, however, and this seems to be the first time that such a relationship has been demonstrated. Further, weak links may stabilize food webs, and can thus be dynamically important even if rare (McCann, 2000).

Seasonal variations in the Broadstone webs showed how the number of connections among species varied over time for a standard unit of effort (i.e., all species and links that were detected in 30 quadrats). As such, comparing connectance webs among sampling occasions provided insight into real patterns in the frequency of interactions per unit area, whereas comparisons with other, less exhaustively sampled, systems should be treated with caution because of autocorrelations between sampling artefacts and web size.

B. Trivariate Relationships, Ontogenetic Shifts, and Seasonally Quantified Webs

Trivariate relationships between body size, abundance, and web structure—similar to those reported recently by Cohen et al. (2003) in the Tuesday Lake food web—were apparent in the Broadstone Stream web, despite the latter having a more truncated size spectrum that spanned only six orders of magnitude in body mass, compared with twelve orders in Tuesday Lake (which contained fish). On average, prey were about ten times smaller than their predators in Broadstone, a generally smaller size disparity than in Tuesday Lake, where predators were often two or three orders of magnitude larger than their prey. However, the maximal size difference between predators and prey in both webs was about six orders of magnitude, suggesting an upper bound to the predator-prey size ratio. In both Broadstone Stream and Tuesday Lake, the vast majority of links represented the consumption of smaller, more abundant prey by larger, rarer predators. These "rules" were broken in a few instances in both systems, with predators sometimes feeding on prey less abundant than themselves. However, there were very few instances where predators fed on prey that were (on average) larger than themselves (18% of observed links), and there was only one link in the Broadstone web where a predator fed on a prey species that was both larger *and* rarer than itself (<1% of observed links). These exceptions can be ascribed to seasonal shifts in the relative body sizes of predators and their prey (described below), revealing the importance of including both temporal and ontogenetic data when assessing food web structure (see also Woodward and Hildrew, 2002b).

After quantifying our seasonal connectance webs, it was clear that the webs were far simpler than implied by the presence-absence data: a few species and links accounted for most of the trophic interactions in the web, whether measured in terms of numbers or biomass of prey ingested. However, because of the negative correlation between body size and abundance, species and links that dominated numerically were often relatively unimportant in terms of biomass, and *vice versa*. The biomass webs represented snapshots of the major pathways of organic matter (energy) flow. Conversely, the abundance webs highlighted the major population dynamics: most individuals that were eaten were small and contributed relatively little to energy flux. Energy flow and population dynamics can be important in different ways in structuring communities (Power, 1990; Wootton, 1997; Hall *et al.*, 2000), and the construction of either biomass or abundance webs without the other would tend to obscure one or other of these processes.

In systems where prey availability is low, such as acid streams, generalist feeding is probably advantageous and this will inevitably result in highly interconnected food webs (Woodward and Hildrew, 2002a). Indeed, the predators in Broadstone were extreme trophic generalists that ate virtually any prey item smaller than themselves. However, this generalism was overemphasized in the connectance web. *Quantitatively*, some prey were clearly overrepresented in predator diets compared with the benthos or the diets of other carnivores; this was largely due to seasonal and ontogenetic differences in size-related handling constraints, which generated differential susceptibility to predators among species and over time. For instance, within the predator guild, cannibalism and mutual predation were particularly prevalent only when generations overlapped, resulting in seasonal "*ontogenetic reversals*" (after Polis *et al.*, 1989) in trophic status that were driven by changes in the relative size of predators and prey (see also Woodward and Hildrew, 2002b). Such feeding loops, once thought to be rare (Pimm, 1982; Cohen and Newman, 1985), now appear to be common in nature, as suggested by the recent "niche models" of food web structure (Warren,

1996; Williams and Martinez, 2000). However, because these situations only arose at certain times of the year, and since only the smallest individuals—which are also those least likely to survive (Hildrew et al., 2004)—were eaten, the effects of ontogenetic reversals were probably relatively weak on average when compared with "top-down" effects. The seasonal and ontogenetic changes in the size-spectrum of the web meant that although 18% of the links were from larger to smaller species, not surprisingly *none* of these individual links represented ingestion of a larger individual by a smaller individual.

During summer and autumn, the web was dominated by a profusion of small detritivorous chironomids that were consumed mostly by the small predators; by winter and spring, the web was dominated by large predators and prey (mostly stoneflies). Consequently, the number, magnitude, and distribution of ingestion rates varied seasonally, with complexity and consumption per unit area (Tables 1 and 4) peaking during summer and declining progressively over time, with many links being broken as prey outgrew their potential predators. Inevitably, the smaller predators were those most affected and the tanypods became increasingly detritivorous as prey availability declined (Smith and Smock, 1992), a trait that would also presumably reduce the risk of intraguild predation while foraging actively for increasingly scarce animal prey (Woodward and Hildrew, 2002b).

The seasonal decline in invertebrate abundance in Broadstone following the summer has been previously attributed to predation because ingestion rates are high and detritus is not limiting to the prey assemblage (Hildrew and Townsend, 1982; Dobson and Hildrew, 1992). Reduced ingestion rates in spring, when prey are rarer, might reflect an increase in the relative availability of physical refugia (Hildrew and Townsend, 1977; Townsend and Hildrew, 1979; Woodward and Hildrew, 2002d), which could potentially stabilize the food web by weakening links (e.g., McCann, 2000). Indeed, consumption rates of the Broadstone predators, which are limited by encounter rate, are markedly lower than their potential (Speirs et al., 2000; Woodward and Hildrew, 2002c). During summer, prey abundance and mobility are both at their peak (Hildrew and Townsend, 1976; Winterbottom et al., 1997) and small individuals dominate the size spectrum (Woodward and Hildrew, 2002b). This would, in theory, maximize both prey availability and predation, as is also suggested by the strong depletion of prey by predators in field experiments at this time of year (Woodward and Hildrew, 2002d). Occasional strong pulses of predation can also occur at other times, such as during spates, when predators and prey are concentrated in flow refugia (Lancaster, 1996). Thus, predator impacts can be highly variable, both seasonally and over much shorter temporal scales.

C. Food Web Topology and Interaction Strength

Recent food web models have shown that complexity (i.e., many species and/ or high connectance, after May, 1972) can enhance web stability if most links are weak, leading ecologists to reassess the established paradigm that complex webs are unstable (May, 1972, 1973; Polis, 1998; McCann, 2000). Unfortunately, interaction strength (*sensu stricto*, May, 1973) is notoriously difficult to measure in real systems, although approximations may be made from empirical and experimental evidence to provide insight into the relative importance of different links (Berlow *et al.*, 2004).

So, accepting these limitations, can we say anything about interaction strengths in the Broadstone food web? There are several independent lines of evidence that lend support to our estimates of link strength (as I/P) from the quantified food web, which suggest that most interactions might be "weak," particularly over intergenerational scales. For instance, at the base of the web, detritus is superabundant; also, links between this dominant basal resource and primary consumers are donor-controlled, suggesting an abundance of weak links, at least between these two trophic levels (Dobson and Hildrew, 1992). Although predatory links do not necessarily follow the same pattern, the Broadstone predators consumed prey biomass equivalent to 65% total benthic production (excluding *Pisidium* sp., which were not eaten), leaving a relatively large portion available for other losses (e.g., disease, drift, and the production of adults). These estimates of consumption are somewhat lower than reported elsewhere (e.g., >100% reported by Allen, 1951; 94% reported by Smith and Smock, 1992), but most other studies have ignored meiofaunal production (but see Stead *et al.*, 2005) and our estimate rises to 80% if the temporary and permanent meiofauna are excluded.

Our estimates of total secondary production, including the permanent meiofauna, in Broadstone (total, 5.22 g m^{-2} y^{-1}; insects only, 3.61 gm^{-2} y^{-1}) are very similar to those reported by Stead *et al.* (2005) from their study in the neighboring Lone Oak stream (total, 4.48 g m^{-2} y^{-1}; insects only, 1.93 gm^{-2} y^{-1}). Further, the predator guild accounted for almost exactly the same proportion of macrofaunal production in both streams (33% in Broadstone, 34% in Lone Oak). Total meiofaunal production, including macrofaunal species within the temporary meiofauna, however, accounted for 52% of the total in Lone Oak but only 19% in Broadstone, primarily because the small chironomids and oligochaetes that dominated the temporary meiofauna in Lone Oak were rarer in Broadstone. Clearly, meiofauna have the potential to contribute significantly to production in other systems, and the widespread omission of this portion of the community size-spectrum might account for some of the unexpectedly high estimates of consumption reported. Predator production in Broadstone was

$1.53 \, \text{g m}^{-2} \, \text{y}^{-1}$, similar to the $1.73 \, \text{g m}^{-2} \, \text{y}^{-1}$ reported by Smith and Smock (1992), but lower than that in the Ogeechee River ($14.35 \, \text{g m}^{-2} \, \text{y}^{1}$), where water temperature was sufficiently high to permit multivoltinism (Benke and Wallace, 1997). In all of these systems, a few prey species supported the majority of predator production, suggesting the predators might have strong effects on only a small fraction of the community. Energy flux, however, does not necessarily equate with interaction strength, unless expressed as the proportion of prey production that is ingested. In Broadstone, our estimates of I/P suggested that only a few links were strong, and that 63% of them accounted for <10% of annual prey production.

In terms of the proportion of annual production that was ingested, the smaller, more 'r-selected' species, which had no access to upper size-refugia, suffered the strongest predation pressure in Broadstone, suggesting that we might be in a position to start identifying species traits (e.g., body size and its associated correlates) that determine the distribution of interaction strength. Some researchers, however, have argued that only experimental manipulations can provide true measures of interaction strength (e.g., Paine, 1992). Although it is logistically impossible to manipulate any more than a tiny proportion of the links within almost any natural food web, field enclosure/ exclosure manipulations of the larger predators in Broadstone have also revealed that only a few prey species were strongly depleted, whereas most others were relatively unaffected (Lancaster et al., 1991; Woodward and Hildrew, 2002d).

All of the lines of evidence cited above are based entirely on short-term (i.e., intragenerational) interactions among the benthic larvae. Many mathematical food web models are, however, constructed using intergenerational population dynamics (e.g., May, 1973; McCann et al., 1998), which are extremely difficult, or even impossible, to measure and parameterize in most natural systems. Some researchers have addressed this by examining interactions among short-lived protists in microcosms under different food web configurations (e.g., Petchey, 2000) but, although they can provide undoubtedly valuable insight into the potential importance of interactions within small food web "modules," such studies can also lack realism because they consider artificial assemblages of species that interact in very simple and homogenous environments (usually glass bottles) with little or no physical refugia or environmental disturbance.

Broadstone Stream, however, is unusual in that we have empirical and experimental data that span a broad range of temporal and spatial scales and degrees of realism. These studies suggest that, although predation can be intense between recruitment periods (Hildrew and Townsend, 1982), it does not appear to destabilize individual prey populations at the intergenerational scale (Speirs *et al.*, 2000). Indeed, long-term data suggest that the entire larval assemblage as a whole is extremely persistent with very little interannual variation, even over several decades, which are equivalent to tens of generations for most taxa (Woodward *et al.*, 2002). The constancy in the composition of the food web may be due to compensatory responses to predation, whereby alternative feeding paths are used at low prey densities (Speirs *et al.*, 2000). This could, in theory, weaken the strength of interactions, thereby increasing the stability of the food web (e.g., McCann, 2000). The system is also resilient to physical perturbations, with rapid recovery following large flood events (Lancaster and Hildrew, 1993). Further, most of the members of the food web are very fecund, and relatively few adults may be required to reset the next generation (Wilcock *et al.*, 2001).

A recent large-scale, intergenerational manipulation of one of the dominant large predators, *Sialis fuliginosa*, which is densely connected within the food web, revealed strongly stabilizing density-dependent mortality, which was ascribed to predation on the early life stages (Hildrew *et al.*, 2004). Following experimental reductions or increases in recruitment of >90% across 150 m stretches of the stream, the effects on population size were only transient, such that they persisted for no more than a few months and did not carry over to subsequent generations (Hildrew *et al.*, 2004). The dynamics of the adult and eggs of most aquatic insects are still poorly understood, but might provide the key to understanding how freshwater communities are able to persist over long time scales, even when competitive and predatory interactions among the aquatic larvae may seem intense (Woodward and Hildrew, 2002a). The predictable seasonality of Broadstone Stream might permit the vast majority of the secondary production of the small and supposedly more "vulnerable" prey to be eaten and yet still allow sufficient prey to survive each year to repopulate the benthos. The current mismatch between the temporal (and spatial) scales at which models are constructed and empirical data are collected has, to date, hindered the advancement of food web ecology in freshwaters and elsewhere, and more large-scale studies are urgently required (Woodward and Hildrew, 2002a).

D. Limitations and Future Directions

Inevitably, there exist numerous sources of potential error in the construction of quantified food webs. For instance, estimates of production and ingestion may be influenced by external subsidies of drifting and terrestrial prey, in addition to vertical colonization from the hyporheos. Fortunately, these confounding effects are likely to be relatively small in Broadstone because the hyporheic zone is very shallow (usually <5 cm), drift is low due to the sluggish nature of the stream (Lancaster and Hildrew, 1993), and the rarity of terrestrial insects in predator guts suggested that they contributed little to energy flux. Similarly, the soft-bodied permanent meiofauna, although eaten by the predators (Schmid-Araya et al., 2002a,b), were unlikely to contribute significantly to total energy flux because of their small size and low production per unit area. For instance, these taxa accounted for only 3% of total production in the neighbouring Lone Oak stream (cf. Stead et al., 2005). Though energetically trivial, however, food web links to the meiofauna could play a strong structuring role by contributing to the early survival of larger predators.

Stable isotope analysis (SIA) offers a potential alternative method of measuring long-term assimilation (a_{ji} in the Jacobian matrix) directly, but it lacks the taxonomic precision of gut contents analysis. The quantitative contributions of iron bacteria and algae to the Broadstone web remain largely uncertain, but an earlier study showed that terrestrial detritus dominated the $\delta^{13}C$ signature of L. nigra, N. pictetii, S. fuliginosa and P. conspersa (Winterbourn et al., 1986). Because these four taxa alone accounted for about 30% of the secondary production in Broadstone, and the only grazer we found was an extremely rare mayfly, any chemosynthetic production by iron bacteria flocs and any conventional primary production by the impoverished algal biofilms in the stream were likely to represent relatively trivial basal inputs. The vast majority (>99%) of the energy flux in the web must rest on the processing of CPOM by detritivores (mostly stoneflies) and FPOM consumption by deposit feeders (mostly chironomids and oligochaetes).

It has been suggested that the considerable effort required to increase the sample of well-described connectance webs might be better directed toward studying processes (e.g., energy flux, ingestion rates) rather than patterns (e.g., connectance) (Hall and Raffaelli, 1993; Benke and Wallace, 1997; Woodward and Hildrew, 2002a). Although the quantified and semi-quantified webs that have been published in the last decade (e.g., de Ruiter et al. 1995; Raffaelli and Hall, 1996) have often emphasized the limitations of qualitative webs, important advances have been made recently in the study of the topological properties of food webs (e.g., Warren, 1996; Williams and Martinez, 2000; Cohen et al., 2003). Food web architecture and dynamics are clearly linked, as we can see in the Broadstone web and in other systems (e.g., Emmerson and Raffaelli, 2004), and the dynamic and static approaches are therefore likely to complement one another, rather than necessarily serving to provide contrasting viewpoints.

The current hindrance to the development of the field is, undoubtedly, however, the shortage of detailed quantified webs that allow us to explore the potential relationships between dynamics and structure. As more, and better quantified, food webs emerge, the search for such generalities (or the lack of them) will become easier. A major challenge for food web ecologists is to produce webs that are standardized sufficiently to allow meaningful comparisons across systems. Producing yield-effort curves, quantifying webs for numbers or biomass consumed per unit area (or volume) per unit time, and expressing ingestion relative to production, for instance, will facilitate such comparisons, which are imperative if we are to parameterize and validate models with real, empirical data (Cohen *et al.*, 1993a; McCann 2000).

Despite this current lack of standardization, however, some surprisingly robust similarities appear to be emerging from some of the better-described webs, especially in relation to the potential structuring role of body size (e.g., Cohen *et al.*, 1993b; Warren, 1996; Martinez and Williams, 2000). Of particular relevance in the light of results from the current study are some of the trivariate relationships between body size, abundance and web topology reported by Cohen *et al.* (2003) for the Tuesday Lake food web, which also hold true for Broadstone Stream. These suggest that food web structure might be bound by a set of rules, potentially relating to energetic constraints. Perhaps equally as intriguing is the suggestion that, at least in Broadstone Stream, there are clear *multivariate* relationships between suites of species traits, food web topology, and interaction strength, derived from body size and its correlates (e.g., P/B ratio, abundance, diet width). These relationships now need to be investigated in a range of different systems if we are to assess their generality.

ACKNOWLEDGMENTS

Financial support for this research project was provided by a Natural Environment Research Council Studentship Grant awarded to G. Woodward while studying at Queen Mary University of London. We would like to thank the Conservators of the Ashdown Forest, the numerous people who helped with fieldwork, Jenny Schmid-Araya for providing additional meiofaunal data, and Tracey Stead for providing data on secondary production in Lone Oak Stream. We also wish to thank Joel Cohen and Dan Reuman for their insightful and helpful comments, which greatly improved the manuscript.

Appendix 1 Generation times for the dominant Broadstone Stream taxa

Order	Taxon	Equivalent Broadstone taxon	Generation time (days)		Mean flight period (days)	Data source
			min	max		
Odonata	Cordulegaster boltonii	C. boltonii	1,095	1,418	42	Woodward, 1999
Megaloptera	Sialis fuliginosa	S. fuliginosa	730*	730	28†	*Speirs et al., 2000
Trichoptera	Plectrocnemia conspersa	Plectrocnemia conspersa	365*	365*	24.5†	*Speirs et al., 2000; †Hildrew, pers. comm.
Trichoptera	Potamophylax cingulatus	P. cingulatus; Adicella reducta	365	365	24.5	Waters, 1977
Plecoptera	Plecoptera	Siphonoperla torrentium, Nemurella pictetii	365*	365*	31.5†	*Hynes, 1977; †Petersen, pers. comm.
Plecoptera	Leuctra hippopus	Leuctra hippopus	365*	365*	31.5†	*Zwick, pers. comm.; †Petersen, pers. comm.
Plecoptera	Leuctra nigra	Leuctra nigra	365*	547.5*	31.5†	*Zwick, pers. comm.; †Petersen, pers. comm.
Ephemeroptera	Paraleptophlebia spp.	Paraleptophlebia submarginata	365	365	24.5	Humpesch, pers. comm.
Diptera	Zavrelymia melanura	Zavrelymia barbatipes	365	365	<7	Morgan, 1980
Diptera	Tanypodinae	Trissopelopia longimana; Macropelopia nebulosa	365	365	<7	Schmid, pers. comm.
Diptera	Chironomous anthracinus	Prodiamesa olivaceae	730	730	<7	Waters, 1977
Diptera	Micropsectra	M. bidentata	365	365	<7	Schmid, pers. comm.
Diptera	Stempellinella	Brillia modesta	365	365	<7	Schmid, pers. comm.
Diptera	Heterotrissocladius marcidus	H. marcidus	365	365	<7	Morgan, 1980
Diptera	Polypedilum spp.	P. albicorne grp.	182.5	365	<7	Schmid, pers. comm.
Simuliidae	Simuliidae	Simulium spp.	121.7	365		Waters, 1977
Amphipoda	Gammarus lacustris	Niphargus aquilex	182.5	182.5	—	Morgan, 1980
Isopoda	Asellus aquaticus	A. meridianus	730	730	—	Waters, 1977
Oligochaeta	Enchytraeidae	Oligochaeta	365	365	—	Bird, 1982
Mollusca	Pisidium crassum	Pisidium spp.	547.5	547.5	—	Morgan, 1980

REFERENCES

Allen, K.R. (1951) The Horokiwi Stream: A study of a trout population. *New Zealand Marine Dep. Fish. Bull.* **10–10a**, 1–231.

Benke, A.C. (1993) Concepts and patterns of invertebrate production in running waters. *Internationale Vereinigung für Theoretische und Angewandte Limnologie, Verhandlungen.* **25**, 15–38.

Benke, A.C. and Wallace, J.B. (1997) Trophic basis of production among riverine caddisflies: Implications for food web analysis. *Ecology* **78**, 1132–1145.

Benke, A.C., Wallace, J.B., Harrison, J.W. and Koebel, J.W. (2001) Food web quantification using secondary production analysis: Predaceous invertebrates of the sang habitat in a subtropical river. *Freshwater Biol.* **46**, 329–346.

Berlow, E.L., Neutel, A.M., Cohen, J.E., de Ruiter, P., Ebenman, B., Emmerson, M., Fox, J.W., Jansen, V.A.A., Jones, J.I., Kokkoris, G.D., Logofet, D.O., McKane, A.J., Montoya, J.M. and Petchey, O. (2004). Interaction strengths in food webs: Issues and opportunities. *J. Anim. Ecol.* **73**, 585–598.

Bird, G.J. (1982) Distribution, life cycle and population dynamics of the aquatic enchytraeid *Propappus volki* (Oligochaeta) in an English chalkstream. *Holarctic Ecol.* **5**, 67–75.

Closs, G.P. and Lake, P.S. (1994) Spatial and temporal variation in the structure of an intermittent stream food-web. *Ecol. Monogr.* **64**, 1–21.

Cohen, J.E. (1978) Food webs and niche space. Monographs in Population Biology. No. 11. Princeton University Press.

Cohen, J.E. and Newman, C.M. (1985) A stochastic theory of community food webs. I. Models and aggregated data. *Proc. Royal Soc. London B* **224**, 421–448.

Cohen, J.E. (1989) Food webs and community structure. In: *Perspectives in Ecological Theory* (Ed. by J. Roughgarden, R.M. May and S. Levin), pp. 181–202. Princeton University Press, Princeton, NJ.

Cohen, J.E., Jonsson, T. and Carpenter, S.R. (2003) Ecological community description using food web, species abundance, and body-size. *Proc. Natl. Acad. Sci. USA* **100**, 1781–1786.

Cohen, J.E., Pimm, S.L., Yodzis, P. and Saldaña, J. (1993b) Body sizes of animal predators and animal prey in food webs. *J. Anim. Ecol.* **62**, 67–78.

Cohen, J.E., Beaver, R.A., Cousins, S.H., DeAngelis, D.A., Goldwasser, L., Heong, K.L., Holt, R.D., Kohn, A.J., Lawton, J.H., Martinez, N., O'Malley, R., Page, L.M., Patten, B.C., Pimm, S.L., Polis, G.A., Rejmánek, M., Schoener, T.W., Schoenly, K., Sprules, W.G., Teal, J.M., Ulanowicz, R.E., Warren, P.H., Wilbur, H.M. and Yodzis, P. (1993a) Improving food webs. *Ecology* **74**, 252–258.

de Ruiter, P.C., Neutel, A.M. and Moore, J.C. (1995) Energetics, patterns of interaction strengths, and stability in real ecosystems. *Science* **269**, 1257–1260.

Dobson, M. and Hildrew, A.G. (1992) A test of resource limitation among shredding detritivores in low order streams in southern England. *J. Anim. Ecol.* **61**, 69–77.

Elton, C. S. (1927) *Animal ecology.* Sedgwick and Jackson, London.

Elser, J.J. and Urabe, J. (1999) The stoichiometry of consumer-driven nutrient recycling: Theory, observations and consequences. *Ecology* **80**, 735–751.

Emmerson, M.C. and Raffaelli, D.G. (2004) Predator-prey body size, interaction strength and the stability of a real food web. *J. Anim. Ecol.* **73**, 399–409.

Goldwasser, L. and Roughgarden, J. (1997) Sampling effects and the estimation of food web properties. *Ecology* **78**, 41–54.

GraphPad Software Inc. (2000) *GraphPad Prism Version 3.0.* GraphPad Software Inc., 5755 Oberlin Drive #110, San Diego, CA 92121, USA.

Hall, R.O., Wallace, J.B. and Eggert, S.L. (2000) Organic matter flow in stream food webs with reduced detrital resource base. *Ecology* **81**, 3445–3463.

Hall, S.J. and Raffaelli, D. (1993) Food webs: Theory and reality. *Adv. Ecol. Res.* **24**, 187–239.

Hildrew, A.G. (1992) Food webs and species interactions. In: *The Rivers Handbook* (Ed. by P. Calow and G.E. Petts), pp. 309–330. Blackwell Sciences, Oxford.

Hildrew, A.G. and Townsend, C.R. (1976) The distribution of two predators and their prey in an iron-rich stream. *J. Anim. Ecol.* **45**, 41–57.

Hildrew, A.G. and Townsend, C.R. (1977) The influence of substrate on the functional response of *Plectrocnemia conspersa* (Curtis) larvae (Trichoptera: Polycentropodidae). *Oecologia* **31**, 21–26.

Hildrew, A.G. and Townsend, C.R. (1982) Predators and prey in a patchy environment: A freshwater study. *J. Anim. Ecol.* **51**, 797–815.

Hildrew, A.G., Townsend, C.R. and Hasham, A. (1985) The predatory Chironomidae of an iron-rich stream: Feeding ecology and food web structure. *Ecol. Entomol.* **10**, 403–413.

Hildrew, A.G., Woodward, G. Winterbottom, J.H. Orton, S. (2004) Strong density-dependence in a predatory insect: Larger scale experiments in a stream. *J. Anim. Ecol.* **73**, 448–458.

Hynes, H.B.N. (1977) A key to the adults and nymphs of the British Stoneflies (Plecoptera). Titus Wilson & Son, Kendal, UK.

Lancaster, J. (1996) Scaling the effects of predation and disturbance in a patchy environment. *Oecologia* **107**, 321–331.

Lancaster, J. and Hildrew, A.G. (1993) Flow refugia and the microdistribution of lotic macroinvertebrates. *J. N. Am. Benthol. Soc.* **12**, 385–393.

Lancaster, J. and Robertson, A.L. (1995) Microcrustacean prey and macroinvertebrate predators in a stream food web. *Freshwater Biol.* **34**, 123–134.

Lancaster, J., Hildrew, A.G. and Townsend, C.R. (1991) Invertebrate predation on patchy and mobile prey in streams. *J. Anim. Ecol.* **60**, 625–641.

Ledger, M.E. (1997) *Grazing of biofilm by invertebrates in streams of contrasting pH.* PhD Thesis. Queen Mary and Westfield College, University of London.

MacArthur, R. H. (1955) Fluctuations of animal populations and a measure of community stability. *Ecology* **36**, 533–536.

Martinez, N.D. (1991) Artifacts or attributes? Effects of resolution on the Little Rock Lake food web *Ecol. Monogr.* **61**, 367–392.

Martinez, N.D., Hawkins, B.A., Dawah, H.A. and Feifarek, B.P. (1999) Effects of sampling effort on characterization of food-web structure. *Ecology* **80**, 1044–1055.

May, R.M. (1972) Will a large complex system be stable? *Nature* **238**, 413–414.

May, R.M. (1973) *Stability and Complexity in Model Ecosystems.* (Princetown University Press).

McCann, K.S. (2000) The diversity-stability debate. *Nature* **405**, 228–233.

McCann, K., Hastings, A. and Huxel, G.R. (1998) Weak trophic interactions and the balance of nature. *Nature* **395**, 794–798.

Morgan, N.C. (1980) Secondary production. In: *The Functioning of Freshwater Ecosystems* (Ed. by E. D. LeCren and Lowe-McConnell),Cambridge University Press, Cambridge, U.K.

Paine, R.T. (1988) On food webs: Road maps of interactions or the grist for theoretical development? *Ecology* **69**, 1648–1654.

Paine, R.T. (1992) Food web analysis through field measurements of per capita interaction strength. *Nature* 355, 73–75.

Petchey, O.L. (2000) Prey diversity, prey composition, and predator population stability in experimental microcosms. *J. Anim. Ecol.* 69, 874–882.

Pimm, S. L. (1980) Properties of food webs. *Ecology* 61, 219–225.

Pimm, S.L. (1982) *Food webs.* Chapman and Hall, New York.

Polis, G.A. (1991) Complex trophic interactions in deserts: An empirical critique of food web theory. *Am. Nat.* 138, 123–155.

Polis, G.A. (1994) Food webs, trophic cascades and community structure. *Austral. J. Ecol.* 19, 121–136.

Polis, G.A. (1998) Stability is woven by complex webs. *Nature* 395:, 744–745.

Polis, G.A., Myers, C.A. and Holt, R.D. (1989) The ecology and evolution of intraguild predation: Potential competitors that eat each other. *Ann. Rev. Ecol. Systemat.* 20, 297–330.

Power, M.E. (1990) Effects of fish in river food webs. *Science* 250, 811–814.

Raffaelli, D. and Hall, S.J. (1996) Assessing the relative importance of trophic links in food webs. In: *Food Webs: Integration of Patterns and Dynamics* (Ed. by G.A. Polis and K.O. Winemiller), pp. 185–1991. Chapman and Hall, New York.

Rundle, S.D. (1988) *The micro-arthropods of some southern English streams.* PhD thesis. Queen Mary and Westfield College, (University of London).

Schmid, P. E., Tokeshi, M. and Schmid-Araya, J. M. (2000) Relationship between population density and body size in stream communities. *Science* 289, 1557–1560.

Schmid-Araya, J.M., Hildrew, A.G., Robertson, A., Schmid, P.E. and Winterbottom, J.H. (2002a) The importance of meiofauna in food webs: Evidence from an acid stream. *Ecology* 83, 1271–1285.

Schmid-Araya, J.M., Schmid, P.E., Robertson, A., Winterbottom, J.H., Gjerlv, C. and Hildrew, A.G. (2002b) Connectance in stream food webs. *J. Anim. Ecol.* 71, 1062.

Slansky, F. and Scriber, J.M. (1982) Selected bibliography and summary of quantitative food utilization by immature insects. *Bull. Entomol. Soc. Am.* 28, 43–55.

Smith, L.C. and Smock, L.A. (1992) Ecology of invertebrate predators in a Coastal Plain stream. *Freshwater Biol.* 28, 319–329.

Smock, L.A. and Roeding, C.E. (1986) The trophic basis of production of the macroinvertebrate community of a southeastern USA blackwater stream. *Holaretic Ecol.* 9, 165–174.

Speirs, D.C., Gurney, W.S.C., Winterbottom, J.H. and Hildrew, A.G. (2000) Long-term demographic balance in the Broadstone Stream insect community. *J. Anim. Ecol.* 69, 45–58.

Stead, T.K., Schmid-Araya, J.M. and Hildrew, A.G. (in press) Meiofauna and the secondary production of a stream metafoam community. *Limnol. Oceanogr.* 50, 398–403.

Strayer, D. and Likens, G.E. (1986) An energy budget for the zoobenthos of Mirror Lake, New Hampshire. *Ecology* 67, 303–313.

Sugihara, G., Bersier, L. F. and Schoenly, K. (1997) Effects of taxonomic and trophic aggregation on food web properties. *Oecologia* 112, 272–284.

Tavares-Cromar, A.F. and Williams, D.D. (1996) The importance of temporal resolution in food web analysis – evidence from a detritus-based stream. *Ecol. Monogr.* 66, 91–113.

Thompson, R.M. and Townsend, C.R. (1999) The effect of seasonal variation on the community structure and food web attributes of two streams: Implications for food web science. *Oikos* **87**, 75–88.

Tokeshi, M. (1999) *Species Coexistence: Ecological and Evolutionary Perspectives.* Blackwell Science.

Townsend, C.R. and Hildrew, A.G. (1977) Predation strategy and resource utilization by *Plectrocnemia conspersa* (Curtis) (Trichoptera: Polycentropodidae). Proceedings of the 2nd International Symposium on Trichoptera, Junk, The Hague.

Townsend, C.R. and Hildrew, A.G. (1979) Resource partitioning by two freshwater invertebrate predators with contrasting foraging strategies. *J. Ani. Ecol.* **48**, 909–920.

Townsend, C.R., Thompson, R.M., McIntosh, A.R., Kilroy, C., Edwards, E. and Scarsbrook, M. (1998) Disturbance, resource supply, and food web architecture in streams. *Ecol. Lett.* **1**, 200–209.

Warren, P.H. (1989) Spatial and temporal variation in the structure of a freshwater food web. *Oikos* **55**, 299–311.

Warren, P.H. (1996) Structural constraints on food web assembly. In: *Aspects of the Genesis and Maintenance of Biological Diversity* (Ed. by M.E. Hochberg, J. Clobert and R. Barbault), pp. 142–161. Oxford University Press, Oxford, U.K.

Warren, P.H. and Lawton, J.H. (1987) Invertebrate predator-prey body-size relationships: An explanation of upper triangularity in food webs and patterns in food web structure. *Oecologia* **74**, 231–235.

Waters, T.F. (1977) Secondary production in inland waters. *Adv. Ecol. Res.* **10**, 91–164.

Wilcock, H.R., Hildrew, A.G. and Nichols, R.A. (2001) Genetic differentiation of a European caddisfly: Past and present gene flow among fragmented larval habitats. *Mol. Ecol.* **10**, 1821–1834.

Williams, R.J. and Martinez, N.D. (2000) Simple rules yield complex food webs. *Nature* **404**, 180–183.

Winemiller, K.O. (1990) Spatial and temporal variation in tropical fish trophic networks. *Ecol. Monogr.* **60**, 331–367.

Winterbottom, J.H., Orton, S.E. and Hildrew, A.G. (1997) Field experiments on the mobility of benthic invertebrates in a southern English stream. *Freshwater Biol.* **38**, 37–47.

Winterbourn, M.J., Rounick, J.S. and Hildrew, A.G. (1986) Patterns of carbon resource utilization by benthic invertebrates in two British river systems. *Archiv für Hydrobiologie* **107**, 349–361.

Woodward, G. and Hildrew, A.G. (2001) Invasion of a stream food web by a new top predator. *J. Ani. Ecol.* **70**, 273–288.

Woodward, G. and Hildrew, A.G. (2002a) Food web structure in riverine landscapes. *Freshwater Biol.* **47**, 777–798.

Woodward, G. and Hildrew, A.G. (2002b) Body-size determinants of niche overlap and intraguild predation within a complex food web. *J. Ani. Ecol.* **71**, 1063–1074.

Woodward, G. and Hildrew, A.G. (2002c) Differential vulnerability of prey to an invading top predator: Integrating field surveys and laboratory experiments. *Ecol. Entomol.* **27**, 732–744.

Woodward, G. and Hildrew, A.G. (2002d) The impact of a sit-and-wait predator: Separating consumption and prey emigration. *Oikos* **99**, 409–418.

Woodward, G., Jones, J.I. and Hildrew, A.G. (2002) Community persistence in Broadstone Stream (U.K.) over three decades. *Freshwater Biol.* **47**, 1419–1435.

Woodward, G., Thompson, R., Townsend, C.R. and Hildrew, A.G. (in press) Pattern and process in food webs: Evidence from running waters. Chapter 16. In *Aquatic Food Webs: An Ecosystem Approach*. (Ed. by A. Belgrano, U. Scharler, J. Dunne and B. Ulanowicz). Cambridge University Press.

Wootton, J.T. (1997) Estimates and tests of *per capita* interaction strength: Diet abundance and impact of intertidally foraging birds. *Ecol. Monogr.* **67**, 45–64.

Wootton, J. T., Parker, M. S. and Power, M. E. (1996) The effect of disturbance on river food webs. *Science* **273**, 1558–1561.

Yodzis, P. (1988) The indeterminacy of ecological interactions as perceived through perturbation experiments. *Ecology* **69**, 508–515.

Yodzis, P. (1998) Local trophodynamics and the interaction of marine mammals and fisheries in the Benguela ecosystem. *J. Ani. Ecol.* **67**, 635–658.

Estimating Relative Energy Fluxes Using the Food Web, Species Abundance, and Body Size

DANIEL C. REUMAN AND JOEL E. COHEN

I. SUMMARY

Given the food web, mean body sizes, and numerical abundances of species in an ecological community, four new models to estimate the relative flux of energy along any pair of links were developed. The models were tested using the data collected by Stephen R. Carpenter and colleagues in Tuesday Lake,

ADVANCES IN ECOLOGICAL RESEARCH VOL. 36
0065-2504/05 $35.00

Michigan, to describe the pelagic food web together with mean body mass (M) and population density (N) of each species. In the metabolic action model, flux was proportional to the product of prey population production times predator population consumption, using allometric formulas for these quantities. This model tested marginally better than the other models and was more easily visualized and applied. Two other models were based on the same allometric formulas, and the fourth was based on an allometric relationship of Emmerson and Raffaelli (2004) between predator impact on prey and the ratio of predator and prey body mass. A new graphical summary of a food web took the $\log(M)$ versus $\log(N)$ plot of Cohen, Jonsson and Carpenter (2003) and added equiproduction and equiconsumption lines, making it possible to visualize species M and N data, trophic data, allometric data, and relative flux data under any of the four models, all from a single plot. The flux models were used to compare several definitions of trophic height; some definitions were more likely than others to correspond to methods of measuring trophic height based on stable isotope analysis. The flux models were also used to develop an ecosystem sampling theory that associated p-values to statements that a given trophic link did not occur in a system. This theory may assist in choosing ecosystems for study that are likely to yield the highest-quality data with the least sampling effort.

II. INTRODUCTION

This report proposes, evaluates, and applies some methods of estimating relative energy fluxes through the trophic links of a community food web, given the average body mass (M) and the numerical abundance per unit of habitat (N) of each species in the web. Previous efforts to estimate fluxes in food webs based on demographic and metabolic data include Moore *et al.* (1993), deRuiter *et al.* (1995), Rott and Godfray (2000), Ulanowicz (1984), Bersier *et al.* (2002), and others.

The combination of food web data with species' M and N data—hereafter called an (M, N)-web—has been explored by Cohen *et al.* (2003), Jonsson *et al.* (2005), and Reuman and Cohen (2004). Those studies and the current study used data collected by Stephen R. Carpenter and colleagues from the community in the nonlittoral epilimnion in Tuesday Lake, Michigan, a small temperate lake that is further described in the Data section below. To our knowledge, Tuesday Lake is the only system with complete, published data on the community food web and the mean body mass and numerical abundance of each species.

Four new models of relative flux through the trophic links of an (M, N)-web were developed in this study, and the models were illustrated using the data of Tuesday Lake. Three models were based on standard allometric formulas of population production and population consumption (Peters, 1983). The first, called the metabolic action (MA) model, set flux through a link proportional to the product of the production of the prey times the consumption of the predator. The fourth flux model was based on an empirical allometric formula from Emmerson and Raffaelli (2004) that related numbers of prey eaten by a predator per unit time to the body mass ratio of predator to prey.

Since direct, empirical measurements of the fluxes in Tuesday Lake were not available, the models were tested indirectly. The first testing method computed, for each intermediate species, the ratio of the total estimated flux into that species divided by the total estimated flux out of that species. Models were judged on their ability to produce ratios greater than 1. Ratios less than 1 were considered unrealistic because they indicated more estimated energy flux out of a species than into it. The second testing method considered the models' ability to estimate fluxes that agreed with allometric estimates of population production and population consumption for all species simultaneously. The MA model performed slightly better than the others on these tests, but its victory was not decisive enough to discard the other models. The main weakness of this study was its inability to compare model-estimated fluxes to empirically measured fluxes, which were unavailable for Tuesday Lake. It may be possible to use the unpublished data of the Broadstone Stream ecosystem (Woodward et al., 2005) and the Ythan Estuary system (Emmerson and Raffaelli, 2004) to test the present models directly.

The MA model was more simply defined, and more easily visualized and applied, than the other models. Starting from the food web plot of Cohen et al. (2003) and Jonsson et al. (2005) on $\log(M)$ (vertical) versus $\log(N)$ (horizontal) axes, this study added equiproduction and equiconsumption lines using the standard allometric formulas of Peters (1983) for population production and consumption. These lines had slopes of $-1/\alpha$ and $-1/\beta$, where α and β are the exponents of M in the allometric formulas for production and consumption, respectively. The strength of flux under the MA model could be easily visualized using these lines. The resulting single plot summarized many aspects of the food web data: body masses, numerical abundances, trophic relations, population production and consumption, and estimated fluxes.

The MA and other models were also applied to a flux-based definition of trophic level (Adams et al., 1983; Winemiller, 1990). These definitions of trophic level gave values that were on average less than the chain-length-based definitions of Cohen and Luczak (1992), Cohen et al. (2003), Jonsson et al. (2005), and Reuman and Cohen (2004). The flux-based definitions would probably produce values more similar to the stable isotope analysis measurements of trophic height of Jennings et al. (2002a,b), and Post (2002).

The flux models were also used to create a model of sampling effort. Assuming sampling methods with certain properties, this model associates a p-value with a claim that a given link did not occur in a web, or that, if it did occur, its flux was less than a certain value. If this model was tested and verified experimentally, it could be used to attach p-values to statistical descriptions of food web topology and (M, N)-web structure. The model could be a new tool for understanding how food web structure varies with varying sampling intensity. The model may also be useful in identifying which ecological systems can be sampled for relatively complete community food webs with minimal sampling effort.

Laboratory biologists have a tradition of choosing a few model organisms in which to study general phenomena. Some of these organisms are carefully chosen for the ease with which they can be manipulated or for the ability to generalize the results of study. In recent years, several model food webs and some model (M, N)-webs have emerged (including, but not limited to, Tuesday Lake, Ythan Estuary, Broadstone Stream, and Little Rock Lake). These webs and others may be the current food web theorists' analog to the laboratory biologists' *E. coli, C. elegans, Drosophila*, zebrafish, and mouse. The data on some of these systems were gathered expressly to completely document a community food web. The data of others were gathered with other goals in mind. To our knowledge, the completeness of food web information that can be expected from a given sampling effort has not been analyzed mathematically before. To produce better data on model ecosystems, such analysis should be combined with the usual considerations of the practicalities of observation and sampling.

Flux estimates have been used in other studies to generate Lotka-Volterra coefficients and to address questions of stability (Moore *et al.*, 1993; Neutel *et al.*, 2002; Emmerson and Raffaelli, 2004). We declined to do this because a flux estimate f_{ij} can be used to generate the corresponding Lotka-Volterra coefficient α_{ij}, but estimating the coefficients α_{ji} or α_{ii} would require making additional tenuous assumptions.

III. FLUX ESTIMATION METHODS: DEFINITIONS AND THEORY OF EVALUATION

A. Notation, Definitions, and Assumptions

The following were taken as given: a list of S species, $S \geq 2$; the predation matrix $W = (w_{ij})$, where $w_{ij} = 1$ if species j eats species i, and $w_{ij} = 0$ otherwise; the average body mass M_i and the numerical abundance N_i per unit habitat of species i. These data were taken as independent of time and space, representing either a steady state or an average of fluctuating states.

The (possibly null) sets of resources and consumers of each species k were defined as $R_k = \{i : w_{ik} = 1\}$ and $Q_k = \{j : w_{kj} = 1\}$. The number of species in R_k was called the generality of species k. The number of species in Q_k was called the vulnerability of species k (Schoener, 1989). Species k was called a consumer if R_k was not empty, and was called a resource if Q_k was not empty. An intermediate species i was one such that both R_i and Q_i were nonempty. An isolated species i was one such that both R_i and Q_i were empty. We assumed that no species was isolated and that the web had a single connected component; if the web had multiple connected components, our methods could be applied to each connected component one at a time.

The outputs of the models in this paper were estimates of the relative flux matrix $F = (f_{ij})$, where f_{ij} was the (average or steady state) flow of energy per unit time from species i to species j, expressed as a dimensionless fraction of all energy fluxes measured in units of calories per unit of time and per unit of habitat (surface area or volume).

Allometric assumptions were: for each species i, the population production P_i and population consumption C_i (in energy units) were approximated by the allometric functions

$$P_i = pN_iM_i^{\alpha} \tag{1}$$

$$C_i = cN_iM_i^{\beta} \tag{2}$$

where p, c, α, and β are all positive constants independent of i, and $\alpha < 1$ and $\beta < 1$ (Peters, 1983). These allometric assumptions implied that, in the plane with horizontal axis log numerical abundance ($\log(N)$) and with vertical axis log body mass ($\log(M)$), the locus of points with constant population production P is a straight line with slope $-1/\alpha$ and the locus of points with constant population consumption C is a straight line with slope $-1/\beta$. (To prove this, let $pNM^{\alpha} = k_1$. Then $\log N + \alpha \log M = k_2$, so $\log M = k_3 - 1/\alpha \log N$. The argument for constant population consumption is similar.)

B. Methods of Estimating Fluxes

We analyzed five methods of estimating fluxes.

Method 0 was an equal flux model (EF). All fluxes were taken to equal $1/L$, where L was the total number of links in the web.

Method 1 was a metabolic action model (MA). Let

$$f1_{ij} = \frac{P_iC_j}{\underset{\substack{\text{trophic} \\ \text{links } (g,h)}}{\sum} P_gC_h} \tag{3}$$

where the sum is over all prey-predator species pairs (g = prey species, h = predator species). The flux from i to j was set proportional to the product of the population production of i and the population consumption of j. This assumption is similar to mass-action laws used in chemistry and in the Lotka-Volterra equations, but concentrations, biomasses or population densities were replaced here by estimates of population production and population consumption. This assumption differs notably from mass action laws based on the biomasses of consumer or resource species, as steady fluxes proportional to biomasses for species of different body sizes could be unsustainable if production scaled less than linearly with body size.

Method 2 was the consumer control model (CC). For each consumer species j with (non-null) resource set R_j, let

$$f2_{ij} = \left(\frac{P_i}{\sum_{g \in R_j} P_g} \right) \left(\frac{C_j}{\sum_{\text{consumers } h} C_h} \right) \qquad (4)$$

The flux into consumer j was set by the population consumption of j, and was distributed over the resources of consumer j in proportion to the population production of each of its resource species.

Method 3 was the resource control model (RC). For each resource species i with (non-null) consumer set Q_i, let

$$f3_{ij} = \left(\frac{P_i}{\sum_{\text{resources } g} P_g} \right) \left(\frac{C_j}{\sum_{h \in Q_i} C_h} \right) \qquad (5)$$

The flux out of resource i was set by the population production of i, and was distributed among the consumers of resource i in proportion to the population consumption of each of its consumer species.

Method 4 was the body mass ratio model (BR). Emmerson and Raffaelli (2004) inferred that in the Ythan estuary a power law relationship holds between per capita interaction strength of a predator j on its prey i, and the ratio of the predator's body size to the prey's body size:

$$I_{ij} = \lambda \left(\frac{M_j}{M_i} \right)^{\gamma} \qquad (6)$$

Emmerson and Raffaelli estimated γ near 0.66. This study used 0.66 exactly. The interaction strengths measured by Emmerson and Raffaelli were equivalent to the coefficients of the quadratic terms in the Lotka-Volterra equations with numerical abundance (not biomass abundance) as the variables. According to these equations, the rate of change of the abundance of resource species i due to species j was

$$\frac{dN_i}{dt} = I_{ij}N_iN_j \tag{7}$$

Our flux, f_{ij}, was a flux of energy proportional to $M_i dN_i/dt$ under the assumption that all species have the same caloric value per unit mass. We combined Eqs (6), (7), and this proportionality relation to obtain the estimate

$$f4_{ij} = \lambda(N_iM_i^{1-\gamma})(N_jM_j^{\gamma}) \tag{8}$$

We chose the value of λ so that the sum of all fluxes in the web was 1.

These methods shared several properties. The sum over all trophic links of all fluxes was 1, using any method. All of the relative flux estimates were dimensionless numbers. Empirical measurement of the absolute flux of any trophic link would identify the multiplier from which the theoretical estimates of relative flux along all remaining links could be converted to estimates of absolute flux. In addition, the relative flux along any trophic link under models MA, RC, and CC was independent of the constants p and c. Finally, the flux formulas in these three models could also be used given any positive P_i and C_j. The expressions for P_i and C_j need not necessarily be allometric formulas.

C. Evaluating the Methods: Theory

The relative flux estimates were evaluated using several tests.

1. Input-Output Ratio Test: Theory

For each intermediate species k in the Tuesday Lake system, the quantity

$$\lambda_k = \frac{\sum\limits_{i \in R_k} f_{ik}}{\sum\limits_{j \in Q_k} f_{kj}} \tag{9}$$

was calculated. This ratio was the sum of the fluxes of energy flowing into species k divided by the sum of the fluxes of energy flowing out of species k. Thus λ_k was expected to approximate the reciprocal of the ecological efficiency (Phillipson, 1966). Values were expected to be distributed around 10 when all species were considered. Values were expected to be higher for warm-blooded species and lower for cold-blooded species. If warm-blooded species generally occurred higher in a food web than cold-blooded species (barring parasites), then λ_k was expected to increase with body mass. If the sum of fluxes flowing into a species k equaled the allometric population consumption of species k, and the sum of the fluxes flowing out of species k equaled the allometric population production of that species, then

$$\lambda_k = C_k/P_k = \frac{c}{p}M_k^{\beta-\alpha} \qquad (10)$$

If $\alpha = \beta$, then a plot of log λ_k versus $\log(M_k)$ should have been flat. If $\beta = 0.75$ and $\alpha = 0.66$, then a plot of log λ_k versus $\log(M)$ should have been linear with the slightly positive slope $\beta - \alpha = 0.09$.

The input-output ratio was used to evaluate all methods of estimating fluxes, and the results are presented below.

2. Crosscheck Test: Theory

The aim of the crosscheck test is to check how nearly the estimated fluxes f_{ij} in Tuesday Lake satisfied the assumptions that $P_i = pN_iM_i^\alpha$ and $C_i = cN_iM_i^\beta$, where p and c are independent of i. The method required the computation of four vectors: the allometric production vector P_{allo}, the allometric consumption vector C_{allo}, the flux production vector P_{flux}, and the flux consumption vector C_{flux}. Specifically,

1. $P_{allo} = (P_1/p, \ldots, P_R/p)$, where R was the number of resources in the web, so P_{allo} had i^{th} component $N_iM_i^\alpha$, which was independent of p;
2. $C_{allo} = (C_1/c, \ldots, C_Q/c)$, where Q was the number of consumers in the web, so C_{allo} had i^{th} component $N_iM_i^\beta$, which was independent of c;
3. P_{flux} had i^{th} component equal to $\sum_{j\in Q_i}f_{ij}$, where i ranged over the resource species (this was the vector of estimated total fluxes out of each resource species);
4. C_{flux} had j^{th} component equal to $\sum_{i\in R_j}f_{ij}$, where j ranged over the consumer species (this was the vector of estimated total fluxes into each consumer species).

If the flux estimates were in perfect agreement with the allometric assumptions, then it would be possible to find constants π and χ such that

$$\pi P_{allo} = P_{flux} \qquad (11)$$

$$\chi C_{allo} = C_{flux} \qquad (12)$$

Fluxes estimated by the CC model were guaranteed to satisfy $\chi C_{allo} = C_{flux}$ for some χ, but not guaranteed to satisfy $\pi P_{allo} = P_{flux}$ for some π. Fluxes estimated by the RC model were guaranteed to satisfy $\pi P_{allo} = P_{flux}$ for some π, but not guaranteed to satisfy $\chi C_{allo} = C_{flux}$ for some χ. Fluxes estimated by the MA model or the BR model were not guaranteed to satisfy either equation.

When the above equations were not satisfied perfectly, we could estimate π and χ by treating Eqs (11) and (12) as linear regression equations constrained to pass through the origin, that is, with zero y-intercept, with unknown slope coefficients π and χ:

$$P_{flux} = \pi P_{allo} + \varepsilon_1 \tag{13}$$

$$C_{flux} = \chi C_{allo} + \varepsilon_2 \tag{14}$$

To see how well the above equations were satisfied, we plotted $\log(P_{allo})$ (on the vertical axis) versus $\log(P_{flux})$ (on the horizontal axis) and $\log(C_{allo})$ versus $\log(C_{flux})$. Then multiplicative scaling of the allometric vector became vertical translation of the data points, and multiplicative scaling of the flux vector became horizontal translation. Neither change affected the residuals of the data from the line of slope 1 which best fitted the points (still in log-log coordinates). We measured the quality of the fit of such a line by means of the standard deviation of these residuals. If the same analysis were repeated with P_{flux} on the vertical axis and P_{allo} on the horizontal axis (or C_{flux} and C_{allo}, respectively), the analogous standard deviation statistic would have been precisely the same as the one just described, because horizontal and vertical residuals to a line of slope 1 are the same. The standard deviation of the residuals to the fitted line of slope 1 was the same as the standard deviation of the residuals to any line of slope 1 because vertical or horizontal translation of the line of slope 1 uniformly adds a constant to all residuals, and this addition does not affect the standard deviation of these residuals, though it changes the mean. So an easily-calculated summary statistic was the standard deviation of the residuals to the line y = x, or std($\log(P_{allo})$-$\log(P_{flux})$) or std($\log(C_{allo})$-$\log(C_{flux})$).

IV. DATA FOR EMPIRICAL EXAMPLE: TUESDAY LAKE, MICHIGAN

Tuesday Lake is a small, mildly acidic lake in Michigan (89°32' W, 46° 13' N). The data used in this study were gathered by Stephen R. Carpenter and colleagues from Tuesday Lake in 1984, and again in 1986. In 1985, the three species of planktivorous fish that lived in the lake were removed, and a single species of piscivorous fish was added. In 1984 and 1986, the fish populations had not previously been exploited and the drainage basin had not previously been developed. The data (given in full by Jonsson et al., 2005) consist of the following for each year (1984 and 1986): a list of species; for each species, its predator species and its prey species (for the body sizes and life stages that were present in the lake); its average body mass M (kg fresh weight per individual); and its numerical abundance N (individuals/m^3 in the epilimnion where the trophic interactions take place). The biomass abundance B (kg/m^3) is M times N. The data represent seasonal averages during summer stratification. Most numerical variables, reported as mean values, were estimated by continuing sampling until the standard error of the mean was less than 10% of the mean. Here only the unlumped web of Tuesday Lake using biological species is described. Data for 1984 and 1986 are treated separately.

V. METHODS

All computations and plotting were done with Matlab version 6.5.0.180913a (R13). Linear regressions were done with the Matlab function "regress". All p-values associated with linear regressions were returned by that function. Normality testing was done with the Jarque-Bera test (Matlab statistics toolbox function "jbtest") and the Lilliefors test (Matlab statistics toolbox function "lillietest"). The Lilliefors test is a simulation-based test that returns p-values only between 0.01 and 0.2. Lilliefors p-values above this range have been reported as >0.2, and values below this range have been reported as <0.01. Both the Lilliefors test and the Jarque-Bera test are composite tests of normality (Lilliefors, 1967; Jarque and Bera, 1987). They are based on qualitatively different aspects of the data, so a set of data was called "normal" only if it passed both tests at the 5% level.

VI. FLUX ESTIMATION METHODS: EVALUATION

Each flux method was tested with the input-output ratio test, the crosscheck test, and other tests using the data of Tuesday Lake. All results below assume $\alpha = 0.75$, $\beta = 0.75$ and $\gamma = 0.66$.

A. Direct Comparison Between Models

Figure 1A plots the log flux of each link according to CC against the log flux of each link according to MA in Tuesday Lake, 1984. Figure 1B does the same for models BR vs. MA in 1984. The log flux of trophic links estimated by each of the five models was also plotted versus the log flux from each of the other models, but the remaining plots are not shown. All plots not involving EF had a general linear trend of slope 1. Plots not involving BR were similar to Fig. 1A, and plots involving BR were similar to Fig. 1B. The sum of the squares of the residuals of these plots from the line y = x (Table 1)

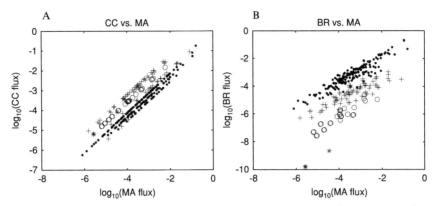

Figure 1 Typical plots of log flux under one model versus log flux under another model, for all trophic links in 1984. Dots denote links with zooplankton predator and phytoplankton prey. The + symbols denote links with zooplankton as both predator and prey. The o symbols denote links with fish as predator and zooplankton as prey. The * symbols denote links with fish as both predator and prey.

Table 1 Sums of squares (rounded to the nearest integer) of residuals for links from the line y = x of log of the flux under the model in the column label versus log of the flux under the model in the row label

	Model			
	MA	RC	CC	BR
1984				
EF	556	608	563	1137
MA	0	60	47	323
RC		0	158	369
CC			0	510
1986				
EF	856	991	428	892
MA	0	14	132	169
RC		0	201	173
CC			0	317

Order of the axes is not important since vertical and horizontal residuals to the line y = x are the same.

showed that MA was similar to RC and CC. RC and CC were also similar to each other, but not as similar as they were similar to MA. None of these models were as similar to BR as they were to each other in 1984, and EF was even more dissimilar from all the other models in both years. In 1986, RC was more similar to BR than it was to CC. Plots with EF on the y-axis had general linear trend of slope 0. Plotting fluxes of any model on the y-axis versus those of the EF model on the x-axis yielded a vertical line, since EF fluxes were all the same.

Exactly linear subtrends of slope 1 occurred in log-log plots of the fluxes under one model versus the fluxes under another model if both models were chosen from among the MA, RC, and CC models (see Fig. 1A). These subtrends can be explained by taking the log of the definitions of flux under these three models:

For MA:

$$\log(f1_{ij}) = \log(P_iC_j) - \log(D) \tag{15}$$

For RC:

$$\log(f3_{ij}) = \log(P_iC_j) - \log\left(\sum_{k\in Q_i} C_k\right) - \log(E) \tag{16}$$

For CC:

$$\log(f2_{ij}) = \log(P_iC_j) - \log\left(\sum_{g\in R_j} P_g\right) - \log(F) \tag{17}$$

where D, E and F are constants. The constant D is the sum over all links of P_iC_j, the constant E is the sum of P_i over all resources in the whole web, and the constant F is the sum of C_i over all consumers in the whole web. When plotting MA versus CC, all links for which the predator species had a fixed set of prey sat on a line of slope 1. When plotting MA versus RC, all links for which the prey species had a fixed set of predators sat on a line of slope 1. When plotting RC against CC, any two links for which the two predators had the same prey set and the two prey had the same predator set sat on a line of slope 1.

The links in both Tuesday Lake food webs were grouped according to whether the predator was a fish (F) or a zooplankton (Z), and whether the prey was a zooplankton or a phytoplankton (P). So all links were classified as (P,Z), (Z,Z), (Z,F) or (F,F) links, where the first letter in the pair gives the group that the prey was in and the second gives the group that the predator was in (Reuman and Cohen, 2004). All of the exactly linear slope 1 subtrends found in plots involving MA, RC, and CC consisted entirely or almost entirely of links from a certain group, as expected, given the characterizations of these exactly linear subtrends found in the previous paragraph (which involved classification by diet and/or predator set).

Exactly linear subtrends were absent in log-log plots of BR-flux versus any one of the MA-, RC-, and CC-fluxes. However, an approximate overall slope-1 trend was visible, and within each group of links there was a clear nonexact linear subtrend. Taking the log of the definition of the BR flux gives:

$$\log(f4_{ij}) = \log(P_iC_j) + \log(M_i^{1-\gamma-\alpha}M_j^{-\beta+\gamma}) + \log(\lambda) \qquad (18)$$

This relation explains the existence of approximate group-based subtrends and the lack of exact subtrends. The variance term (the second one on the right) was not constant for any particular group of links, nor was the difference between this term and the analogous terms in the RC and CC equations. However, the second term in the above equation took a very different distribution of values over links from different groups. Since the two exponents in that term were both negative for the assumed values of α, β and γ, for groups (A, B) where A and B both contain heavy species, we should expect linear subtrends below the overall linear trend. For groups in which A and B were both comparatively light, we should expect linear subtrends that are above the overall trend. These expectations are confirmed in Fig. 1B.

Histograms (not shown) of the flux in the links of the 1984 and 1986 webs under each model except the EF model confirmed the general expectation that a web should have many weak links and few strong links (Paine, 1992; Raffaelli and Hall, 1996; McCann et al., 1998; Kokkoris et al., 1999). Woodward et al. (2005) recently confirmed this phenomenon experimentally. In Tuesday Lake in 1984, for each model except the EF model, the sum of the 14 strongest fluxes under that model (14 of 269 links was a little more than 5%) made up at least 65% of the total flux in all links under that model. The top three fluxes (a little more than 1% of the links) made up at least 29% of the total flux for each model.

Lorenz curves (Fig. 2 for the MA model; Lorenz curves for other models except EF look similar) measure the level of inequality in flux distributions. The horizontal axis of the Lorenz plot shows the cumulative fraction of

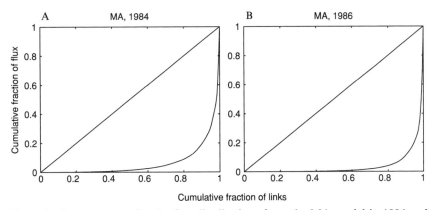

Figure 2 Lorenz curves for the flux distributions from the MA model in 1984 and 1986. The horizontal axis is the cumulative fraction of links, when the links are ranked from lowest to highest flux. The vertical axis is the cumulative fraction of total flux that flows along the links included so far.

Table 2   Gini indices for the flux distributions from each model in each year

Year	MA	RC	CC	BR
1984	0.874	0.871	0.864	0.910
1986	0.925	0.936	0.875	0.892

links, when the links are ranked from lowest to highest flux, while the vertical axis shows the cumulative fraction of total flux that flows along the links included so far. A highly unequal flux distribution would have a Lorenz curve lying far below the line y = x, while a hypothetical flux distribution with all fluxes equal would have Lorenz curve coinciding with the line y = x. The inequality in a flux distribution is quantified by the Gini index, which is twice the area between the Lorenz curve and the line y = x. The Gini index ranges from 0 when all links have equal flux to 1 in the limit when all flux in the system passes along a single trophic link and all remaining links have vanishingly small flux. The Gini indices (Table 2) for the flux distributions of each model were all greater than 0.85. The Gini index and Lorenz curve were not changed when all fluxes were multiplied by any positive constant, and were therefore useful for relative flux distributions.

The log-flux distributions in both years (histograms in Fig. 3) for the MA, RC, and CC models were normal at the 5% significance level, according to the Jarque-Bera and Lilliefors composite tests of normality. The BR distribution was not normal in 1984, but was in 1986. The p-values for these tests are in Table 3.

To summarize, direct comparison of the four models revealed that the MA, RC, and CC models were more similar to each other than they were to the BR model. All models produced very unequal distributions of fluxes with many weak fluxes and a few strong fluxes. Distributions of log-flux were approximately normal for the MA, RC, and CC models in 1984 and 1986, but normal for the BR model only in 1986.

B. Input-Output Ratio Test: Results

The results of applying the input-output ratio test to the five models were as follows.

1. Distributions of Flux Ratios

The log of the input-output flux ratio was computed for each intermediate species for each year and for each model. A species whose only predator was

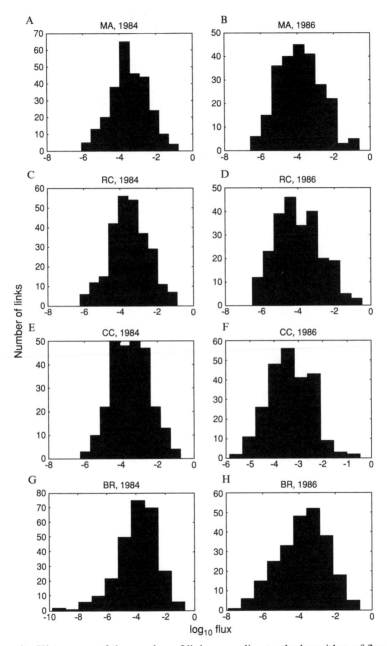

Figure 3   Histograms of the number of links according to the logarithm of flux for all models (except EF) in both years. Distributions are not statistically distinguishable from normal except BR in 1984.

Table 3   Tests of normality of the distribution of the logarithm of flux over links

Species	Year	MA	RC	CC	BR
Strict upper triangular	1984	0.95/>0.2	0.82/>0.2	0.70/>0.2	<0.001/<0.01
	1986	0.20/0.20	0.20/>0.2	0.51/0.05	0.15/>0.2
Upper triangular	1984	0.93/>0.2	0.83/>0.2	0.62/>0.2	<0.001/<0.01
	1986	0.18/0.19	0.16/>0.2	0.82/0.17	0.11/0.13
All	1984	0.95/>0.2	0.85/>0.2	0.59/>0.2	<0.001/<0.01
	1986	0.16/0.09	0.13/>0.2	0.86/0.10	0.10/0.09

Top two rows refer to links in the strict upper triangle in a body mass indexed predation matrix. Middle two rows refer to links in the upper triangle in a body mass indexed predation matrix. Bottom two rows refer to all links. Jarque-Bera (on the left in each cell) and Lilliefors (on the right in each cell) p-values assess normality of log-flux distributions of Tuesday Lake data using each flux model. Low values of p reject lognormality. Only the BR model rejects lognormality, and only in 1984.

Table 4   Minimum and maximum input-output flux ratios for each model in each year, and the number of the 25 intermediate species in 1984 and 21 in 1986 that had input-output flux ratio less than 1

Model	Number of species with flux ratio <1, 1984	Number of species with flux ratio <1, 1986	Minimum flux ratio, 1984	Minimum flux ratio, 1986	Maximum flux ratio, 1984	Maximum flux ratio, 1986
EF	0	0	1.20	1.00	12	23
MA	3	15	0.48	0.16	350	32
RC	3	14	0.50	0.15	156	23
CC	0	2	1.08	1.00	350	19
BR	0	1	1.71	0.67	40940	12035

Cannibalistic species were counted as intermediate.

itself through cannibalism was included as an intermediate species. Ratios < 1 indicated a greater flux out of a species than into it, and were the strongest indicator of fault in a model. Table 4 shows the number of intermediate species with input-output flux ratio <1 for each model and how much less than 1 these ratios were. Maximum ratios are also shown. From plots of log input-output flux ratio versus log body mass for intermediate species (Fig. 4), the distribution of log input-output flux ratios is easily seen by looking only at the ordinate (y-axis value) of each plotted point. An input-output flux ratio <1 appears in Fig. 4 as a log ratio <0.

Although the MA and RC models had many intermediate species with input-output flux ratios less than 1 in 1986, these ratios were rarely much less than 1, and therefore did not represent a serious inaccuracy of the model (Table 4, Fig. 4).

In 1984, the three species with input-output flux ratio less than 1 under the MA model were the same as those with input-output flux ratio less than 1 under the RC model, and these were the only species with ratio less than 1 in

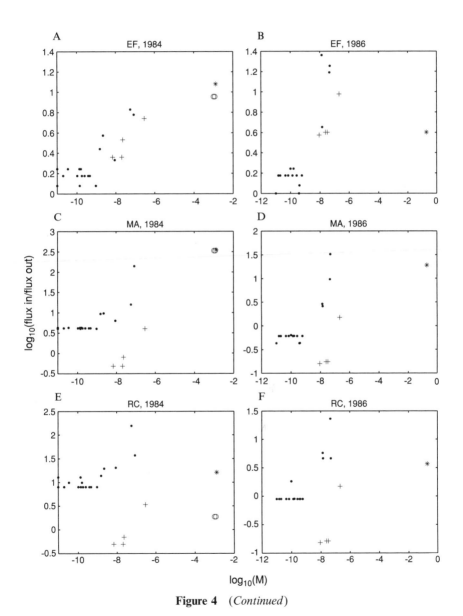

Figure 4   (*Continued*)

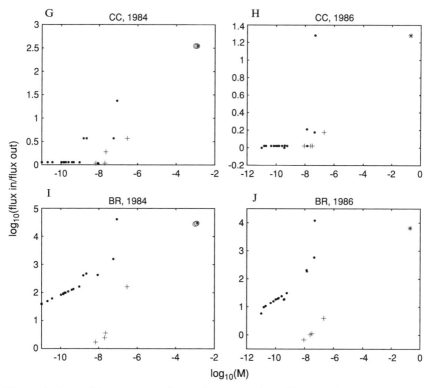

Figure 4 Log of input-output flux ratios versus log of intermediate species body mass for each model in each year. Asterisks (∗) represent cannibalistic fish, ∘ non-cannibalistic fish, + non-fish cannibals, and dots other intermediate species.

any model in that year. These three species, all cannibals, were *Cyclops varians rubellus, Orthocyclops modestus,* and *Tropocyclops prasinus.* They were the only three species from the Cyclopoida order (in the class of Copepods) found in Tuesday Lake in either year. It is plausible that cannibalism occurred between different size classes within each species. These three species were also all present in 1986. If one discounts these species in 1986, then the minimal input-output flux ratios were 1.0000, 0.4314, 0.8903, 0.9981, and 3.9439 for the EF, MA, RC, CC, and BR models, respectively. These values are better than the minima reflected in Table 4, and are either greater than 1 or just slightly less than 1.

The models with the fewest species with input-output flux ratios less than 1 were the BR and CC models, but because flux ratios that were less than 1 for the other models were usually not much less than 1, the input-output flux ratio data did not strongly favor the BR and CC models.

2. *Flux Ratios versus Body Mass*

Input-output flux ratios were plotted against species body mass on log-log scales (Fig. 4). Sometimes cannibalistic species and/or fish were outliers on these plots, so they have been marked separately. Table 5 contains linear regression statistics for these plots, with separate statistics computed without fish and/or cannibalistic species.

One species in 1984 and each of two in 1986 took fewer than 15% and greater than 0% of its prey species from the animal kingdom. None of these were counted as being primarily carnivorous. Any other species that ate any meat at all took at least 95% of its prey species from the animal kingdom. All species that were primarily carnivores were also either fish or cannibals (or both) in both years, so excluding fish and cannibals was the same as excluding all mainly carnivorous species.

The slopes of the regressions in Table 5 were always slightly positive, or not statistically different from zero. This is consistent with the allometrically predicted input-output flux ratios (see the section on Input-output ratio tests: Theory). The EF model had positive regressions in both years, regardless of which outliers were removed. The MA model also had positive regressions unless only fish were removed. The RC model never had slope statistically different from zero unless both fish and cannibals were removed (the only fish in 1986 was also a cannibal, so removing only cannibals was the same as removing fish and cannibals). The CC model always gave positive slopes, except when only fish were removed in 1986. The BR model always gave positive slopes except both years when only fish were removed.

The flux ratio versus body mass trends supported all models because the regressions in the Table 5 were of the correct order of magnitude (theory predicted slope $\beta - \alpha$, so regression slopes should have been between about -1 and 1, which they were). These slopes did not appear to support one model over the others.

C. Flux Differences versus Species Metabolism

The difference between flux in and flux out should equal the amount of metabolic energy consumed by an intermediate species, neglecting energy lost through feces. Log of (flux in minus flux out) was plotted versus the log of allometrically estimated species metabolism using the formula NM^{δ} for metabolism, where $\delta = 0.75$ and 0.80 were both tried (Peters, 1983). It was expected that the line $y = x$ would fit the resulting plot well, but on the contrary there was no visible linear relationship. The noise in both the independent and dependent variables on this plot appeared to overwhelm any pattern that may exist.

Table 5 Linear regression statistics for plots of log(flux in/flux out) versus log(body mass)

	EF	MA	RC	CC	BR
All species					
1984 slope	0.12	0.24	−0.05	0.32	0.34
	(0.09,0.14)	(0.14,0.34)	(−0.15,0.05)	(0.26,0.38)	(0.18,0.50)
r^2	0.83	0.52	0.04	0.84	0.45
p-value	0.00	0.00	0.32	0.00	0.00
1986 slope	0.10	0.16	0.07	0.12	0.25
	(0.03,0.18)	(0.05,0.27)	(−0.04,0.18)	(0.07,0.18)	(0.04,0.45)
r^2	0.31	0.34	0.08	0.53	0.25
p-value	0.01	0.01	0.21	0.00	0.02
No fish					
1984 slope	0.14	0.04	−0.05	0.16	0.16
	(0.10,0.19)	(−0.14,0.22)	(−0.25,0.16)	(0.06,0.25)	(−0.16,0.48)
r^2	0.67	0.01	0.01	0.38	0.05
p-value	0.00	0.65	0.64	0.00	0.32
1986 slope	0.27	0.16	0.08	0.09	0.14
	(0.18,0.35)	(−0.03,0.35)	(−0.11,0.27)	(−0.07,0.18)	(−0.20,0.49)
r^2	0.70	0.15	0.04	0.17	0.04
p-value	0.00	0.09	0.41	0.07	0.40
No cannibals					
1984 slope	0.12	0.27	−0.05	0.33	0.39
	(0.08,0.15)	(0.22,0.32)	(−0.13,0.04)	(0.27,0.39)	(0.31,0.47)
r^2	0.78	0.87	0.08	0.89	0.86
p-value	0.00	0.00	0.24	0.00	0.00
1986 slope	0.33	0.38	0.30	0.15	0.63
	(0.22,0.44)	(0.25,0.52)	(0.20,0.41)	(0.03,0.27)	(0.46,0.81)
r^2	0.74	0.72	0.72	0.33	0.82
p-value	0.00	0.00	0.00	0.02	0.00
No fish or cannibals					
1984 slope	0.16	0.26	0.22	0.22	0.57
	(0.09,0.22)	(0.14,0.37)	(0.12,0.32)	(0.11,0.34)	(0.43,0.72)
r^2	0.64	0.58	0.57	0.53	0.82
p-value	0.00	0.00	0.00	0.00	0.00
1986 slope	0.33	0.38	0.30	0.15	0.63
	(0.22,0.44)	(0.25,0.52)	(0.20,0.41)	(0.03,0.27)	(0.46,0.81)
r^2	0.74	0.72	0.72	0.33	0.82
p-value	0.00	0.00	0.00	0.02	0.00

See Fig. 4. Only intermediate species were included in the first regression. All cannibalistic species were counted as intermediate. The "No fish" and "No cannibals" regressions considered only non-fish and non-cannibalistic intermediate species, respectively. In parentheses are 95% confidence intervals.

D. Crosscheck Test: Results

For each model in each year, $\log(C_{flux})$ was plotted versus $\log(C_{allo})$, and $\log(P_{flux})$ was plotted versus $\log(P_{allo})$. Figure 5 shows the MA and BR plots in 1984. For these plots, the C_{flux}, C_{allo}, P_{flux}, and P_{allo} vectors were multiplicatively normalized (before taking logs) so that each had a Euclidean length of 1. Table 6 has the summary statistic $\text{std}(y_{\text{data}} - x_{\text{data}})$ discussed in the section on Crosscheck test: Theory, including or excluding cannibals and (independently of cannibals) fishes (cannibals and fishes were frequently outliers). The assessment number in the last part of that table, an overall description of each models' performance, is the mean of the nonzero P and C

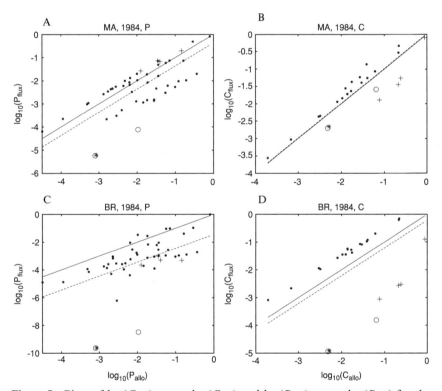

Figure 5 Plots of $\log(C_{flux})$ versus $\log(C_{allo})$ and $\log(P_{flux})$ versus $\log(P_{allo})$ for the MA and BR models in 1984. The proximity of the data on the C plots to a line of slope 1 measures how well the allometric population consumption agreed with total flux into each species. The proximity of the data on the P plots to a line of slope 1 measures how well the allometric population production agreed with total flux out of each species. The solid line is the line y = x, and the dashed line is the least squares line of slope 1. Asterisks (∗) are cannibalistic fish, ○ non-cannibalistic fish, + non-fish cannibals, and dots other species.

Table 6   Assessments of fit of C_{flux} with C_{allo} and of P_{flux} with P_{allo} by means of the standard deviation of the residuals of the data from the line of slope 1 which best fitted the points in log-log coordinates

		EF	MA	RC	CC	BR
All species						
P	1984	0.96	0.75	0.00	1.00	1.54
	1986	1.10	0.32	0.00	0.58	1.08
C	1984	0.76	0.36	0.56	0.00	1.22
	1986	0.80	0.42	0.52	0.00	1.03
No fish						
P	1984	0.98	0.62	0.00	0.96	0.85
	1986	1.11	0.29	0.00	0.57	0.75
C	1984	0.78	0.35	0.57	0.00	0.88
	1986	0.77	0.42	0.52	0.00	0.90
No cannibals						
P	1984	0.99	0.72	0.00	0.99	1.41
	1986	1.12	0.30	0.00	0.57	0.70
C	1984	0.73	0.22	0.40	0.00	0.97
	1986	0.74	0.36	0.42	0.00	0.13
No fish or cannibals						
P	1984	1.01	0.62	0.00	0.97	0.85
	1986	1.12	0.30	0.00	0.57	0.70
C	1984	0.75	0.12	0.33	0.00	0.07
	1986	0.74	0.36	0.42	0.00	0.13
Overall assessment numbers						
All species		0.91	0.46	0.54	0.79	1.22
No fish		0.91	0.42	0.54	0.77	0.84
No cannibals		0.90	0.40	0.41	0.78	0.80
No fish or cannibals		0.90	0.35	0.38	0.77	0.43

The lower the standard deviation, the better the fit. The overall assessment number of each model is the mean of the nonzero P and C statistics from 1984 and 1986. The overall assessments favor the MA model.

table values in 1984 and 1986 for that model. Lower assessment numbers indicated models for which allometric and flux vectors can be better reconciled. The MA model was consistently best by this standard.

In plots of C_{flux} versus C_{allo} and P_{flux} versus P_{allo}, linear subtrends were visible for some of the models. For the EF model, these linear subtrends had a slope of 0. Some of the other models had linear subtrends of slope 1. Fish and cannibals were also sometimes outliers from the general trend on some plots for some models (Fig. 5).

The following equations helped to explain these phenomena. For the EF model,

$$\log(C_{flux,i}) = \log(g_i) - \log(L) \tag{19}$$

$$\log(P_{flux,i}) = \log(v_i) - \log(L) \tag{20}$$

where g_i is the generality of species i, v_i is the vulnerability of species i, and L is the total number of links in the web.

For the BR model,

$$\log(C_{flux,i}) = \log(C_{allo,i}) + \log\left(M_i^{\gamma-\beta}\sum_{j \in R_i} N_j M_j^{1-\gamma}\right) + \log(\lambda) \qquad (21)$$

$$\log(P_{flux,i}) = \log(P_{allo,i}) + \log\left(M_i^{1-\gamma-\alpha}\sum_{j \in Q_i} N_j M_j^{\gamma}\right) + \log(\lambda) \qquad (22)$$

For the MA model,

$$\log(C_{flux,i}) = \log(C_{allo,i}) + \log\left(\sum_{j \in R_i} P_j\right) - \log(D) \qquad (23)$$

$$\log(P_{flux,i}) = \log(P_{allo,i}) + \log\left(\sum_{j \in Q_i} C_j\right) - \log(D) \qquad (24)$$

The second term on the right of (23) is the log of the sum of the productions of the species that i ate, and the second term on the right of (24) is the log of the sum of the consumptions of the species that ate i.

For the RC model,

$$\log(C_{flux,i}) = \log(C_{allo,i}) + \log\left(\sum_{j \in R_i} \frac{P_j}{\sum_{k \in Q_j} C_k}\right) - \log(E) \qquad (25)$$

There is no need for an equation relating $P_{flux,i}$ and $P_{allo,i}$ because the RC model forced them to be equal.

For the CC model,

$$\log(P_{flux,i}) = \log(P_{allo,i}) + \log\left(\sum_{j \in Q_i} \frac{C_j}{\sum_{k \in R_j} P_k}\right) - \log(F) \qquad (26)$$

There is no need for an equation relating $C_{flux,i}$ and $C_{allo,i}$ because the CC model forced them to be equal.

Each of Eqs (21) through (26) has three terms on the right side. The last term is always constant, and the first term is always either $\log(P_{allo,i})$ or $\log(C_{allo,i})$, depending on whether the left side of the equation is $\log(P_{flux,i})$ or $\log(C_{flux,i})$. As a result, there is an underlying linear relationship of slope 1 between $\log(P_{flux,i})$ and $\log(P_{allo,i})$, and between $\log(C_{flux,i})$ and $\log(C_{allo,i})$. This first term will be called the main term. The second term on the right side expresses the variance of the data from the trend. This second term will be called the variance term. In the EF Eqs (19) and (20), the first term on the right side will be called the variance term. The main term is zero for the EF equations because in this case, the underlying trend is a slope 0 trend.

The variance terms are useful in understanding observed linear subtrends, and observed tendencies of certain classes of species to lie far from the linear trend of slope 1 (or slope 0, in the case of the EF model). Two species for which the variance terms were equal sat together on a line of slope exactly 1 (or for the EF model, exactly 0). So exactly linear subtrends in the data arose from classes of species that all shared the same variance term. In the case of the EF model, such species all had the same generality or vulnerability. There were only 11 distinct nonzero generalities in the 1984 web, and nine in the 1986 web. There were eight distinct nonzero vulnerabilities in the 1984 web and seven in the 1986 web. Each of these corresponds to an exactly flat subtrend in the EF plots.

The variance term in the consumption equation (23) for the MA model represents the log of the sum of the productions of the species that species i ate. If two species had identical diets, their variance terms in that equation were the same, and they sat on a line of slope exactly 1 in the plot of $\log(C_{flux})$ versus $\log(C_{allo})$. There were only 14 distinct columns in the 1984 predation matrix, and 11 in the 1986 predation matrix. Because many species shared the same diet, exactly linear subtrends of slope 1 appeared in the MA plots of $\log(C_{flux})$ versus $\log(C_{allo})$ (e.g., Fig. 5B).

In the production Eq. (24) for the MA model, the variance term represents the log of the sum of the consumptions of the species that ate species i. There were 16 distinct rows in the 1984 predation matrix and 13 in the 1986 predation matrix. Because many species shared the same predator set, exactly linear subtrends of slope 1 appeared in the MA plots of $\log(P_{flux})$ versus $\log(P_{allo})$ (e.g., Fig. 5A).

The variance term in the RC Eq. (25) was also the same for two species that had the same prey set, and the variance term in the CC Eq. (26) was the same for two species that had the same predator set. This explains the appearance of exactly linear subtrends of slope 1 in those plots.

The variance terms in the BR equation are not the same for species that had the same prey set or predator set. As expected, the BR plots have no exactly linear subtrends.

Species that were eaten only by fish were outliers on the BR model production plots in 1984 (Fig. 5C) and 1986. The variance terms in the BR model Eqs (21) and (22) explain why. If we assume that all species in Tuesday Lake had the same biomass abundance B (which is a rough but reasonable approximation for present purposes), then we can write the variance terms as

$$\log\left(M_i^{\gamma-\beta} \sum_{j \in R_i} \frac{B}{M_j^{\gamma}} \right) \tag{27}$$

and

$$\log\left(M_i^{1-\gamma-\alpha} \sum_{j\in Q_i} \frac{B}{M_j^{1-\gamma}} \right)$$ (28)

in the consumption (21) and production (22) equations, respectively. If species i was eaten only by fish, then the production variance term was very negative. So we expect species that were eaten only by fish to lie significantly below the overall linear trend of slope 1 on the production plot. The only species that were eaten only by fish were the fish themselves, and one other species in each year (a single species that survived from 1984 to 1986). The fish and this other species deviated from the general trend more than any other species in both years (Fig. 5C, D).

Fish and cannibals were outliers on the BR model consumption plots in 1984 and 1986, and they lay below the overall trend (Fig. 5C, D). These were the same species, in both years, as those with diets that consisted predominantly of meat. For the other models also, though to a lesser extent, the species that lay significantly below the overall linear trend on $\log(C_{flux})$ versus $\log(C_{allo})$ plots had predominantly meat diets. For the MA, RC, and BR models in each year, Fig. 6 plots species' residuals from the line $y = x$ on $\log(C_{flux})$ versus $\log(C_{allo})$ axes, versus the percent of the species' diet that consisted of meat (as calculated using the flux model in question). Species with more meat in the diet generally had more negative residuals. Why?

Consumption of zooplankton (meat) may have been more beneficial to the growth and reproduction of a consumer than consumption of phytoplankton, if zooplankton contained a higher proportion of fat than phytoplankton. Fat has more calories per unit mass than protein or carbohydrate. However, for stoichiometric reasons, zooplankton consumption may have been more beneficial to consumer growth and reproduction even if zooplankton fat content was no higher than that of phytoplankton. The zooplankton of Tuesday Lake may have contained limiting growth reagents in greater abundance than the phytoplankton. In either case, one would expect mainly carnivorous species to fall below the general linear trend on $\log(C_{flux})$ versus $\log(C_{allo})$ plots because less consumption (C_{flux}) of richer food was needed to meet fixed allometric requirements (C_{allo}).

A modified method of evaluating the five flux models was considered. Instead of plotting normalized C_{flux} versus normalized C_{allo} on log-log axes, a normalized alternate C_{flux} (called AC_{flux}) versus a normalized C_{allo} was plotted. The new AC_{flux} was the sum of inbound fluxes as calculated by the flux model under study, but with the fluxes coming from nonphytoplankton species multiplied by some fixed "meat benefit ratio" which was greater than or equal to 1. A summary statistic of the quality of the new plots is $std(\log(AC_{flux}) - \log(C_{allo}))$. This summary statistic was plotted as a function

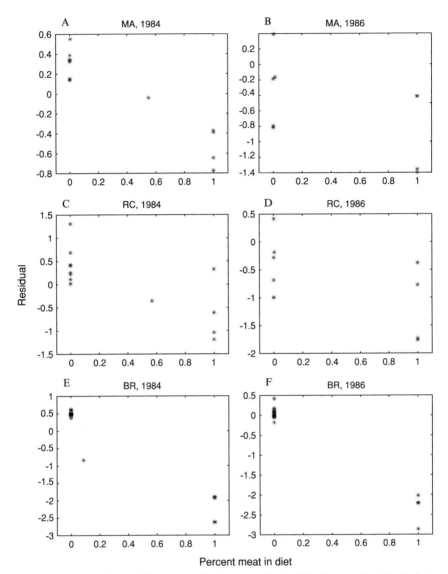

Figure 6 Plots of the residuals from the line y = x in log(C_{flux}) versus log(C_{allo}) plots versus percent of the species' diet that was meat (as calculated using each flux model), for nonbasal species, for models MA, RC and BR in 1984 and 1986.

of the meat benefit ratio, increasing from a meat benefit ratio of 1, for each of the models MA, RC, and BR. For all models in both years, the summary statistic decreased initially as the meat benefit ratio increased from 1, until it reached some minimum after which it increased. Table 7 gives summary

Table 7 Assessments of fit of $\log(C_{flux})$ with $\log(C_{allo})$ and of $\log(AC_{flux})$ with $\log(C_{allo})$[a]

Model	Year	Standard deviation with MBR = 1	MBR for minimal summary stat.	Minimal standard deviation
MA	1984	0.36	5.4	0.15
	1986	0.42	2.3	0.39
RC	1984	0.56	8.3	0.38
	1986	0.52	3.2	0.47
BR	1984	1.22	536	0.20
	1986	1.03	218	0.19

[a]By means of the standard deviation of the residuals of the data from the line of slope 1 which best fitted the points in log-log coordinates. Values in column 5 are for the optimal meat benefit ratios (MBR) shown in column 4.

statistics with meat benefit ratio 1, as well as meat benefit ratios that minimized the summary statistic, and the accompanying minimal statistic. The meat benefit ratio that minimized the summary statistic was called the minimizing meat benefit ratio.

The MA and the RC models had reasonable minimizing meat benefit ratios (between 2.3 and 8.3). The BR did not (its minimizing meat benefit ratio was over 200), but given these unreasonable meat benefit ratios, this model produced good summary statistics, bettered only by the MA model in 1984. The BR model may have improved so much with the implementation of a meat benefit ratio because it was the only model originally posed as a model of biomass flux. The BR model's assumption that all species had the same caloric value per unit mass is precisely the assumption that a nonunit meat benefit ratio seeks to correct. The other models were based on allometric formulas with energy units, and were therefore direct models of energy flux.

Excluding mainly carnivorous species was the same, in both years, as excluding fish and cannibals. The "no fish or cannibals" section of Table 6 shows how well each model performs, considering only species whose diet did not consist mainly of meat. In the two consumption C rows of this part of the table, the BR model outperformed the other models.

VII. APPLICATION: TROPHIC LEVEL AND TROPHIC HEIGHT

Adams *et al.* (1983) proposed and Winemiller (1990) among others used a recursive definition of trophic level τ as follows. Species that ate no other species were assigned trophic level of $\tau = 0$. The trophic level τ_j of any consumer j was defined as

$$\tau_j = 1 + \sum_{i \in R_j} \tau_i F_{ij} \tag{29}$$

where F_{ij} was the fraction of the consumed food of species j consisting of species i. Adams $et\ al.$ (1983) measured the fraction in terms of volume (equivalent to energy under the assumption that all species had the same energy per unit of volume). Using energy flux here, F_{ij} for the Tuesday Lake data was computed from the fluxes f_{ij} by

$$F_{ij} = \frac{f_{ij}}{\sum\limits_{g \in R_j} f_{gj}} \tag{30}$$

The trophic level τ was not constrained to be an integer, and was defined regardless of cannibalism, omnivory, or loops in the food web, although in some of these cases, linear algebraic equations had to be solved.

The fluxes of the five flux models gave five different measures of trophic level. However, the fluxes from the metabolic action and consumer control models gave the same trophic levels for all species. This identity held because if i_1 and i_2 were two prey of species j, then

$$f_{i_1 j}/f_{i_2 j} = P_{i_1}/P_{i_2} \tag{31}$$

for both the MA fluxes and the CC fluxes. So the normalized fluxes F_{ij} in the trophic level equation were the same in both cases. Therefore, for all analyses of trophic level, results were computed for all models except the CC model.

Reuman and Cohen (2004) defined trophic height in a way that did not depend on fluxes, but only on the food web, as follows. The $trophic\ position$ of a species in a food chain was defined to be the number of species below it. (In a recursive definition, species A was said to be $below$ species $B \neq A$ in a food chain if species A was eaten by species B, or if species A was below any species that was below species B.) The $trophic\ height$ (H) of a species was the average trophic position of the species in all food chains to which it belonged, only considering food chains with no repetitions of species. Excluding repetitions of species ruled out chains that went all the way around a loop in the food web (even a loop of length one, i.e., a cannibalistic link). Chains that went any part of the way around a loop were allowed. This definition was the same as one of the definitions in Cohen and Luczak (1992) and Yodzis (1989).

The trophic level of each species was computed under each of the five flux models, and the trophic height of each species was also computed using the method of Yodzis (1989), Reuman and Cohen (2004) and others. Trophic height and trophic level generally increased with increasing species body mass according to any method of calculation.

The trophic height was greater than or equal to the trophic level as measured using the EF or BR flux models, for all species in both years. The trophic height was greater than or equal to the trophic level as measured under the other flux models except for two species in 1984 (the same two species for all three models). This consistent inequality can be understood mathematically in the following way. The trophic height of a species was a weighted average of the trophic heights of its prey, plus one. Prey with more chains entering them from below in the web were weighted more heavily. However, higher trophic height prey tended to have more chains entering them than lower trophic height prey. Thus, the weighted average that produced the trophic height of a species more heavily weighted prey of greater trophic height. This weighting inflated results compared to trophic level, using any one of the flux estimate methods. While trophic level also used a weighted average of the prey of a species, the weighting was based on the percentage of the diet that each prey represented.

In the absence of flux measurements or estimates, one could replace trophic height with trophic level, using the EF model. The assumptions of this model are false, but this method avoids the overestimation problems of trophic height.

The data of Tuesday Lake provided a weak basis for comparing methods of calculating trophic height or level because only a few species in Tuesday Lake had trophic height or level greater than 1 under any method. All basal species had height or level 0 under any method. Basal species were at least half of all species in Tuesday Lake. Species that ate only basal species had height or level equal to 1 under any method. Very few species remained in Tuesday Lake after species of height or levels 0 or 1 were eliminated. Larger webs with more species of height or level greater than 1 are needed to make better comparisons among the methods.

Stable nitrogen isotopes have been widely used to estimate trophic height or level (Peterson and Fry, 1987; Kling *et al.*, 1992; Zanden and Rasmussen, 1999; Post *et al.*, 2000; Post, 2002; Jennings *et al.*, 2002a). The method is based on the fact that the index $\delta^{15}N$ of the ratio of the stable isotopes of nitrogen (see Jennings *et al.*, 2002a for a definition of $\delta^{15}N$) in a predator is approximately 3–4‰ more than the weighted mean of the $\delta^{15}N$ values of its prey species, where the weighting is according to the ease with which the predator absorbs nitrogen from each of its prey species (DeNiro and Epstein, 1981; Minagawa and Wada, 1984; Peterson and Fry, 1987; Post, 2002). Assuming that absorption of nitrogen is proportional to absorption of energy, the mean can be calculated with weighting given by the energy fluxes from each prey to the predator. Stable isotope methods of measuring trophic position were judged to be more likely to correspond closely to trophic level than to trophic height, and such measurements should correspond most closely to the trophic level estimate that is based on the

Table 8 Trophic heights or levels of all nonbasal, nonherbivorous species using six methods of measuring height

Species	Height	EF Level	MA Level	RC Level	CC Level	BR Level
1984						
Cyclops varians rubellus	2.50	2.14	2.67	2.64	2.67	2.15
Daphnia pulex	1.43	1.11	1.00	1.00	1.00	1.00
Orthocyclops modestus	3.20	2.21	3.01	3.05	3.01	2.23
Tropocyclops prasinus	2.50	2.14	2.67	2.64	2.67	2.15
Chaoborus punctipennis	3.60	2.17	2.04	2.08	2.04	1.10
Phoxinus eos	4.17	2.53	3.09	2.93	3.09	2.47
Phoxinus neogaeus	4.17	2.53	3.09	2.93	3.09	2.47
Umbra limi	4.84	2.80	3.13	3.87	3.13	2.47
1986						
Cyclops varians rubellus	3.19	2.23	2.50	2.55	2.50	2.08
Daphnia pulex	1.39	1.10	1.00	1.00	1.00	1.00
Daphnia rosea	1.47	1.13	1.01	1.00	1.01	1.00
Orthocyclops modestus	3.19	2.23	2.50	2.55	2.50	2.08
Tropocyclops prasinus	3.22	2.25	2.55	2.59	2.55	2.09
Chaoborus punctipennis	3.97	2.27	3.75	3.78	3.75	2.38
Micropterus salmoides	4.86	2.79	3.86	3.33	3.86	2.70

most realistic flux model. Therefore, it may be possible to test the flux models using stable isotope analysis, once community-wide (M, N)-food web data are gathered in conjunction with stable isotope analysis data from all or several species in a community.

The trophic heights and levels (under each of the five flux models) of all nonbasal, nonherbivorous species are in Table 8.

VIII. APPLICATION: EQUIPRODUCTION AND EQUICONSUMPTION LINES

Using the allometric formulas for population production in Eq. (1) and population consumption in Eq. (2), the equiproduction and equiconsumption curves on $\log(M)$ (ordinate) versus $\log(N)$ (abscissa) coordinates were lines of slope $-1/\alpha$ and $-1/\beta$, respectively. Including these lines on the food web plot in the plane of $\log(M)$ versus $\log(N)$ (Cohen *et al.*, 2003; Jonsson *et al.*, 2005) made the resulting plot even more powerful for visualizing food webs (Fig. 7B, which assumes $\alpha = \beta = 0.75$, making the equiproduction and equiconsumption lines the same). If $\alpha \neq 1$ and $\beta \neq 1$, the equiproduction and equiconsumption lines do not coincide with the equal biomass lines of slope -1. Hence, there may exist pairs of species i and j such that i has more biomass than j, but j has greater population consumption than i, and similarly for population production.

Under the MA model, the flux along a link was proportional to the product of the population consumption of the predator times the population production of the prey. When equiproduction and equiconsumption lines were added to the $\log(M)$ versus $\log(N)$ plot of the food web, links that had a predator on a high equiconsumption line and a prey on a high equiproduction line had a strong flux under the MA model. For instance, the flux from species 17 (unclassified flagellates) to species 47 (*Chaoborus punctipennis*) in Fig. 7B was very strong under the MA model, while the flux from species 22 (*Chromulina* sp.) to species 39 (*Keratella testudo*) was weak.

Under the CC and MA models, the relative strengths of the fluxes into a consumer were determined by the relative productions of the respective prey. Under either of these models, one could see which of two fluxes into a fixed consumer was stronger (and therefore which of two prey was more important for that consumer) by looking at a $\log(M)$ versus $\log(N)$ plot of the food web, with added equiproduction lines. The more important prey was on a higher equiproduction line. In Fig. 7B, the flux from 47 to 48 (*Phoxinus eos*) was stronger than the flux from 36 (*Holopedium gibberum*) to 48, so species 47 was probably a more important food source for species 48 than species 36 was.

Under the RC and MA models, the relative strengths of fluxes out of a resource were determined by the relative consumption of the respective predators. One could see which fluxes were stronger, and therefore which predators of a given prey consumed the most, using $\log(M)$ versus $\log(N)$ plots with equiconsumption lines. The most consumptive predator was the one on the highest equiconsumption line.

If $\alpha \neq \beta$, the equiproduction and equiconsumption lines of Fig. 7B would no longer coincide, but relative flux strengths could still be visualized in a similar way. Under the BR model, the strength of a flux was proportional to the product of $N_i M_i^{1-\gamma}$ times $N_j M_j^{\gamma}$, for prey i and predator j. Lines of equal $NM^{1-\gamma}$ and NM^{γ} would make possible similar visual comparisons of flux strength.

IX. APPLICATION: ESTIMATING REQUIRED LEVEL OF SAMPLING EFFORT

Cohen *et al.* (1993) suggested that food web data should be accompanied by yield-effort curves, which have units of sampling effort along the x-axis, and either number of observed species or number of observed links along the y-axis. Woodward *et al.* (2005) implemented this suggestion in their study of the Broadstone Stream ecosystem. Their species-yield-effort curves

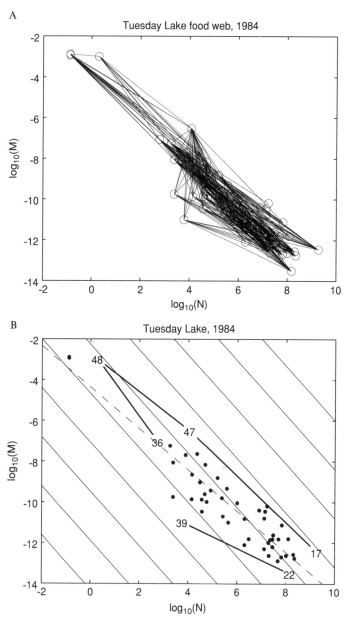

Figure 7 The Tuesday Lake food web in 1984. (A) As plotted by Cohen *et al.* (2003) and Jonsson *et al.* (2005). (B) With added equiproduction and equiconsumption lines (light solid lines), but only selected trophic links (heavy lines that join species numbers). Allometric formulas $P_i = pN_iM_i^{0.75}$ and $C_i = cN_iM_i^{0.75}$ have been assumed, so each equiproduction line coincides with an equiconsumption line. The dashed line

measured effort by the number of examined 25 cm by 25 cm quadrats of habitat. The species-yield-effort curves flattened out with increasing effort, indicating that the listed collection of species would probably not become much more complete with more sampling effort. Woodward *et al.* (2005) presented nine link-yield-effort curves for the nine most common predator species in their ecosystem, with the number of predator guts visually inspected as the measure of effort, and the number of prey species discovered in the guts as the measure of yield. Their link-yield-effort curves did not flatten out, except for some very common and very exhaustively sampled predators (for which hundreds of guts were examined). This finding supported the suspicion that typical reported food webs probably have not documented all links in the system (Martinez, 1991, Martinez *et al.*, 1999). We developed a theory for estimating the probability with which an unobserved link could be said not to exist, or not to represent a flux of more than a given strength. We also developed a theory for estimating the probability of absence of an undetected species.

Woodward *et al.* (2005) captured and counted all species that were larger than a certain size in each sampled 25 cm by 25 cm quadrat. Our model assumed that one unit of species sampling effort (quadrat) was sufficient to observe all species larger than a fixed size that were in a sampled habitat volume or area A. The model ignored species too small to be observed via the sampling method, although such species (e.g., microbes) may be biologically very important. The model also assumed that some version of gut content analysis of predators was used to detect links. Analysis of one gut was considered one unit of link sampling effort. The expected number of individuals of a given species in volume or area A was NA, where N was the numerical abundance (population density) of the species, as before, assuming that the species' presence or absence was independent of the presence or absence of other species in the sampled quadrats. The probability of finding the species in one unit of species-sampling effort was

$$p_s = 1 - P_{NA}(0) \qquad (32)$$

where $P_{NA}(0)$ was the probability of 0 in a Poisson distribution with parameter NA. The probability of not finding the species after n units of sampling effort was therefore $(P_{NA}(0))^n = e^{-NAn}$. This probability was less than a fixed acceptable probability of failure p_f if and only if

---

is a least squares fit (in log-log coordinates) to all the plotted species. Most species are indicated with a dot, but the ones involved in a pictured trophic link are indicated with their species index as given in the appendix of Reuman and Cohen (2004): 17, unclassified flagellates; 22, *Chromulina* sp.; 36, *Holopedium gibberum*; 39, *Keratella testudo*; 47, *Chaoborus punctipennis*; 48, *Phoxinus eos*.

$$n \geq \frac{-\ln(p_f)}{NA} \qquad (33)$$

After n units of sampling effort, each species with density N greater than

$$\frac{-\ln(p)}{nA} \qquad (34)$$

would have been detected with probability at least $1 - p$. If after n units of sampling effort a species was not observed, then with probability $1 - p$ its numerical abundance was less than the quantity in Eq. (34).

Rare species that required many units of sampling effort to detect with reasonable probability were probably also large (because of the negative correlation between size and abundance). A second sampling method could be used to detect the presence of any such species that may have been missed using the first sampling method (e.g., nets with bigger mesh size that could be dragged over a larger area or volume of habitat).

Let V be a predator's gut volume, and let J be its consumption rate in volume units per unit time (which is assumed proportional to M^β, with M the average body mass of the predator, under the assumption that all species had equal energy density). If gut residence time of food particles is proportional to V/J (Kooijman, 2000, p. 81), and if V is proportional to the predator's mass M, then gut residence time is proportional to $M^{1-\beta}$. Let φ_{ij} be the aggregate (over all individuals) absolute energy flux per unit time from species i to species j. This energy flux is proportional to biomass flux under the assumption that all species have roughly the same energy content per unit mass. If prey species i was recognizable for a fixed proportion of the time it spent in its predator's gut, then the average number of individuals of species i that could be recognized in the gut of a randomly chosen individual of species j was proportional to

$$\eta = \frac{\varphi_{ij} M_j^{1-\beta}}{M_i N_j} \qquad (35)$$

Let k denote the constant of proportionality. An italic f will be used to denote a relative flux, and a φ will be used to denote an absolute flux. Then the probability of having found evidence of species i in the gut of an individual of species j was

$$p_l = 1 - P_{k\eta}(0) = 1 - e^{-k\eta} \qquad (36)$$

If n was the number of guts that were examined and p_f was the acceptable probability of not finding a link that exists, then $e^{-kn\eta} \leq p_f$ if and only if

$$n \geq \frac{-\ln(p_f) M_i N_j}{k\varphi_{ij} M_j^{1-\beta}} \qquad (37)$$

Under the MA model, the right side of Eq. (37) became

$$\frac{-\ln(p_f)M_i^{1-\alpha}}{kpcN_iM_j} = \xi \tag{38}$$

If M_i increased, then N_i tended to decrease, and therefore ξ (the required sampling effort for a given probability of finding the link) increased. Modifying M_j had the opposite effect. The links most difficult to detect (in terms of the number of guts that had to be examined) had relatively small consumer M_j and relatively large resource M_i. These same conclusions held for the BR model.

After n units of sampling effort, each link with flux greater than

$$\frac{-\ln(p)M_iN_j}{knM_j^{1-\beta}} \tag{39}$$

would have been detected with probability at least $1 - p$, as a consequence of Eq. (37). If after n units of sampling effort a link had not been observed, then with probability $1 - p$, its flux was less than Eq. (39).

The MA and other proposed flux models are models of relative flux, but Eqs (35)–(39) use absolute flux. If a link from species i to species j was detected, one could estimate from experimental data the probability of having found evidence of species i in a single gut of species j by counting the percentage of guts of species j in which species i was found. Setting this relative frequency equal to the probability in Eq. (36) yields an estimate of $k\varphi_{ij}$. Doing this for all detected fluxes and fitting the estimates to

$$k\varphi_{ij} = kpcN_iN_jM_i^\alpha M_j^\beta \tag{40}$$

yields absolute estimates of $k\varphi_{ij}$ for unsampled or undetected links. In addition, the quality of the fit of the right side of Eq. (40) to the empirical estimates of $k\varphi_{ij}$ would be a valuable test of the MA model (or another model, if one replaces Eq. (40) with the flux definition for another model). Then the sentence that contains Eq. (39) gets replaced with the statement that each link for which the absolute $k\varphi_{ij}$ estimate was greater than

$$\frac{-\ln(p)M_iN_j}{nM_j^{1-\beta}} \tag{41}$$

would have been detected with probability at least $1 - p$ after n units of link-sampling effort. If after n units of sampling effort a link had not been observed, then with probability $1 - p$ its $k\varphi_{ij}$ estimate was less than Eq. (41).

This analysis makes it possible to identify in advance a food web that will have links and species that can be easily detected. Before a link is observed, its predator species must be observed, and predators tend to be rarer than prey. Several predator guts may have to be examined before a given link is detected. So an ecosystem with no extremely rare species, and especially no extremely rare predators, can be more easily studied. A reviewer suggested that species-poor ecosystems may be good examples of systems with few extremely rare species because their species body mass distributions may have shorter upper tails. Tuesday Lake, Broadstone Stream, and Skipwith Pond—three of the most detailed webs currently available—are all relatively species-poor acid systems.

The probability of a link being detected, given the predator's gut, is given by Eq. (36). In a web that is well-suited for study, this probability must be large for all links. Therefore, the web must have large η and large $\log(\eta)$ for all links. Assuming the MA model, taking $\log(\eta)$, throwing out constant terms, and making use of the allometric relationship $\log(N) = -s\log(M) + \kappa$ (Peters, 1983), an easily studied web would have only links with large values of

$$\log(M_j) - (s + 1 - \alpha)\log(M_i) = \log\left(\frac{M_j}{M_i^\varepsilon}\right) \tag{42}$$

where $\varepsilon = s + 1 - \alpha$, which is close to 1. Therefore, the most easily sampled ecosystems will only have links with large predator to prey mass ratios, and no very rare predators.

This condition may be satisfied in an ecosystem with a classification of species into groups of very different body masses, with all members of each group feeding only on members of other groups of smaller body mass. The existence of such a classification would imply a block structure in a body mass indexed predation matrix, but is not equivalent to such a block structure. The Tuesday Lake body mass indexed predation matrix has a block structure in both 1984 and 1986 (Reuman and Cohen, 2004), and the species in Tuesday Lake can be classified into phytoplankton, zooplankton (which feed mainly on the phytoplankton), and fish (which feed mainly on the zooplankton). However, the gap in body mass between zooplankton and phytoplankton is not large enough in Tuesday Lake to ensure that all trophic links have large predator-to-prey body mass ratios. Moreover, some zooplankton feed on other zooplankton of similar size, and some fish feed on other fish. One reviewer suggested that pelagic systems may have exclusively high predator-to-prey body mass ratios because of the common trophic and size separation among the categories of phytoplankton, zooplankton, and fish. Other pelagic systems should be examined to see if it is possible to find a system with a clear size gap between phytoplankton and zooplankton, in which all zooplankton are herbivores, and no fish eat other fish. The same reviewer pointed out that benthic systems in freshwater tend not to have well-separated size and trophic classifications of species, due to the commonness of insects.

A similar contrast between marine systems and some terrestrial systems has been noticed (Pauly *et al.*, 2002). The size spectrum of some pelagic systems can be manipulated for experimental purposes through the use of nets of variable mesh size. The condition of large predator-to-prey body mass ratios in every trophic link may also be satisfied by a community in which all species have dentition and feeding practices that allow them only to eat species much smaller than themselves.

Conclusions similar to Eq. (42) hold for the BR model, but with exponents $1 - \beta + \gamma$ on M_j and $s + \gamma$ on M_i on the right side. Qualitative characterization of easily studied webs using the RC or CC models is more difficult.

The above approach is equally applicable when the link-sampling method is traditional visual gut-content analysis or a more sophisticated and sensitive method of gut content analysis. The only change would be the value of the constant k in Eqs. (36)–(40). Polymerase chain reaction (PCR) has been used to identify mosquito larvae in the gut contents in dragonfly nymphs (Morales *et al.*, 2003) and to distinguish among three prey species in the guts of spiders (Greenstone and Shufran, 2003). To our knowledge, PCR has not been used community-wide to improve identification of prey species in the guts of predators. Use of PCR in this way might increase the interval after predation within which gut contents can still be identified, decreasing the total number of guts that have to be examined to detect a link with a fixed probability of success. Considering that the yield-effort curves for the number of links of Woodward *et al.* (2005) flattened out only for predators for which hundreds of guts were sampled, some such improvement may be essential to gather complete data on any food web of reasonable complexity.

X. APPLICATION: MEAN TRANSFER EFFICIENCIES

Jennings *et al.* (2002b) and Gaedke and Straile (1994) calculated mean transfer efficiencies in the following way. They grouped individual organisms into bins of log body mass (with no regard to the species of the organism). They then created a "production-size-spectrum," which is a histogram with $\log(M)$ bins on the x-axis, and log of the amount of production occurring in the organisms within that weight class on the y-axis. This log production can be computed for each bin as α times the central $\log(M)$ value for that bin plus the log of the number of organisms in that bin. The "central" $\log(M)$ value of a bin is the arithmetic mean of the minimal and maximal $\log(M)$ values included in that bin. The log of the mean transfer efficiency was then the linear-regression slope of the production-size-spectrum times the mean of the logs of the predator-prey body mass ratios. These mean transfer efficiency estimation methods were adapted so that they could be applied to (M, N)-food webs by assuming that all individuals of a given species have log-body-mass equal to log(M) given in the (M, N)-web data. The adapted methods were used to calculate the mean transfer efficiencies in Tuesday Lake.

Given a relative flux model, mean transfer efficiencies were computed for Tuesday Lake by adding the relative fluxes coming out of all intermediate species, and dividing by the sum of the relative fluxes going into the same set of species. The quotient obviated any need for absolute fluxes. Comparison of the outputs of these two methods was used as a test of the relative flux models (Table 9).

To compute mean transfer efficiencies using the adapted methods of Jennings *et al.* (2002b), log(M) bins beginning at -14 in both years and ending at -2 in 1984 and 0 in 1986 were used. These values approximately delimited the values of log(M) in each year. Regression values for the slope of the resulting production-size spectrum depended slightly on the width of the log(M) bins that were used. Table 9 shows the resulting transfer efficiencies for a reasonable range of bin widths. All flux-model transfer efficiencies were comparable with the values obtained using the adapted methods of Jennings *et al.* (2002b), except for the BR model values, which were 1–2 orders of magnitude too small.

The main weakness of the methods of Jennings *et al.* (2002b), as those methods were adapted here to (M, N)-webs (not a shortcoming in the original methods), was that the choice of bin widths could affect the estimate of transfer efficiency. Jennings *et al.* (2002b) worked with a log(M) distribution of individuals, while the current study worked with the distribution of the logarithms of body mass means over species. The latter distribution was

Table 9 Mean transfer efficiencies in Tuesday Lake according to each of several computational methods

	1984	1986
EF flux	0.383	0.325
MA flux	0.257	0.273
RC flux	0.127	0.150
CC flux	0.491	0.629
BR flux	0.008	0.008
bin width 2	0.259	0.387
bin width 1	0.631	0.685
bin width 0.5	0.494	0.646

The methods marked with a flux model take the quotient of the fluxes out of intermediate species by the fluxes into intermediate species. The methods marked by a bin width use the adapted methods of Jennings (2002b), as described in the text.

much coarser and therefore more prone to yield different results with different bin choices. For Tuesday Lake data in both years, for reasonable $\log(M)$ bin widths, the latter distribution had some bins that contained no species. The resulting $-\infty$ log-production values for those bins were ignored for linear regressions of log-production versus $\log(M)$. If the original distributions of individual body masses over each species had been retained in addition to the means of these distributions, these shortcomings could have been remedied.

XI. DISCUSSION

Given the food web, mean body sizes, and numerical abundances of species in an ecological community, the relative flux of energy along any link was estimated in several plausible ways. Previous efforts to estimate fluxes include Moore *et al.* (1993), deRuiter *et al.* (1995), Rott and Godfray (2000), Ulanowicz (1984), Bersier *et al.* (2002), and others. Several new models of the flux of energy were proposed here. Models of relative energetic flux were also models of relative biomass flux if multiplication by a constant sufficed to convert biomass to energy. When resource species differed in their energetic value (Cousins, 2003), the conversion between biomass and energy would be conditional on both the resource and the consumer.

A. Which Model Is Most Plausible

A null model supposed all fluxes equal (EF model). Three models of relative flux were based on allometric relations between mean body mass and population production and population consumption (the MA, RC, and CC models). One model of flux was based on an allometric relation between the rate of consumption and the body mass ratio of predator to prey (the BR model, adapted from Emmerson and Raffaelli (2004)). Lacking direct empirical estimates of fluxes, in this paper we evaluated the relative merits of the models using two indirect methods based on the input-output ratio for each species and the cross-check of predicted fluxes against allometric assumptions.

The input-output ratios under any model in either year were never much less than 1, except for three problematic cannibalistic species from the order Cyclopoida in both years (these species were present in both years, and were the only species from that order). In 1984, no model had input-output ratios less than 1 at all, except for these three species under the MA and RC models. In 1986, only the MA and RC models had any input-output ratios less than 0.99, other than these three problematic cannibals, and these ratios were also not much less than 1. The greater presence of input-output ratios less than 1 in 1986 could be a fault of the MA and RC flux models, or an indication that the system was not at equilibrium in 1986. Since the input-output ratios were only slightly less than 1, these results may also mean nothing. The BR and CC models were the best models (other than EF) for ensuring no input-output ratios less than 1 in Tuesday Lake. Plots of log flux ratios versus log body mass revealed either no trend or an increasing trend, fulfilling the predictions of theory.

The consumer crosscheck test measured the consistency between the population consumption of each consumer species predicted by each model and the population consumption predicted by an allometric formula. The resource crosscheck test similarly compared population production of each resource species to flux out of those species according to a given model. Ignoring any comparison that is guaranteed perfect by definition of the model, we found that the MA model was the most consistent (among the models considered here) with the underlying allometric assumptions, for the Tuesday Lake data. This superiority of the MA model is no guarantee that it will perform better than the other models when absolute or relative fluxes are measured directly, or that it will still be the most consistent model when applied to other (M, N)-food web data sets.

On plots of $\log(C_{flux})$ versus $\log(C_{allo})$, mainly carnivorous species frequently lay below the general slope 1 linear trend, especially under the BR model. This pattern may arise because zooplankton (meat) prey species had a caloric or stoichiometric advantage over phytoplankton prey species. When we multiplied meat fluxes by a meat benefit ratio, the plots improved (i.e., the scatter from a slope 1 linear trend diminished) as the meat benefit ratio increased from 1, for MA, RC, and BR models in both years ($\log(C_{flux})$ versus $\log(C_{allo})$ plots for the CC model were perfect by definition). Meat benefit ratios that maximized the quality of the plots seemed quantitatively plausible (between 2.3 and 8.3) for the MA and RC models, but did not seem quantitatively plausible (218 and 536) for the BR model, although the $\log(C_{flux})$ versus $\log(C_{allo})$ plots with minimizing meat benefit ratios were better for the BR model than for either the RC or MA models. Using the minimizing meat benefit ratio for each model, the mean qualities of $\log(C_{flux})$ versus $\log(C_{allo})$ plots and $\log(P_{flux})$ versus $\log(P_{allo})$ plots were still best for the MA model.

Theoretical predictions of relative fluxes could be converted to theoretical predictions of absolute fluxes if the absolute flux of one or more links were measured empirically. To test directly whether one model was more successful than another would require empirical estimates of the relative or absolute fluxes of at least two links (in addition to the assumed information about the food web, mean body sizes, and numerical abundance of all species in those links).

B. Applications, Implications, and Possible Future Directions

Plausible flux estimates promise a variety of applications. In this study, we used flux estimates to compare six measures of trophic height or trophic level: two measures based on the food web alone (or, for the EF model, the food web plus the assumption of equal fluxes in all links), and four measures based on the flux models proposed here. The measure based on mean lengths of food chains used by Cohen *et al.* (2003), Jonsson *et al.* (2005), Reuman and Cohen (2004), and others probably inflated results compared to plausible stable isotope measures of trophic level. The measures of trophic level based on the flux models would probably be more in tune with stable isotope measures of trophic level.

To see which flux model estimates trophic levels closest to those obtained from stable isotopes, complete (M, N)-food web data would have to be compared with stable isotope measurements of several species. The isotope analysis should be done on species high in a food web, because it is for such species that estimates of trophic height or level differ the most among alternative models. A web with some omnivory should be used, since omnivory gives rise to differences among methods of calculating trophic level and height.

All flux models considered here produced flux distributions that were extremely unequal (except the EF model). The MA, RC, and CC models produced normally distributed log-flux distributions in both years. The BR model did so only in 1986. These results suggested log-normality of fluxes as a testable null hypothesis to quantify the qualitative hypothesis of "many weak, few strong links." Quantitative measures of community flux distribution should be produced for other webs using estimated and (when available) directly measured fluxes. The Lorenz curve and the Gini index are two convenient measures of inequality in flux distributions.

Plausible flux estimates also made possible a theory of the amount of sampling effort needed to detect links and species in a community, with a given probability of success, when using sampling methods with certain properties. The accuracy and usefulness of this sampling theory could be tested on any data that include summary (M, N)-food web data and detailed records of the sampling process, including amounts and timing of sampling effort and the fruits of each unit of sampling effort. The unpublished data of Woodward *et al.* (2005) on the Broadstone Stream ecosystem may contain these details. The sampling theory presented here related the sampling effort expended, the population density of a species to be measured, and the probability of detecting that species. The theory also provided a similar relationship between sampling effort, the strength of flux through a link, and the probability of detecting the link. An experimentally verified theory of sampling effort could be useful for associating levels of certainty with observed statistical regularities in food webs. Such a theory may also be useful in selecting for study an ecosystem that could be sampled with minimal effort to provide comprehensive or near-comprehensive (M, N)-food web data.

Mean trophic transfer efficiencies computed using an adaptation of the methods of Jennings *et al.* (2002b) and Gaedke and Straile (1994) were found to be comparable to mean trophic transfer efficiencies computed using the flux models, except the BR model. Adapted methods had to be used because Tuesday Lake data included only the mean body mass for each species. By contrast, the original methods of Jennings *et al.* (2002b) and Gaedke and Straile (1994) use the masses of each individual organism captured. The adaptation created uncertainty in the resulting mean transfer efficiencies, so that results could not be used as evidence that one flux model is superior to the others, except in the case of the BR, where the disagreement between the two methods was pronounced. Data including body mass measurements for all individual organisms captured would be necessary for more precise comparison, and this comparison may provide a way of distinguishing among flux models.

Comparing the Ecosim and Ecopath fisheries models of Villy Christensen, Carl Walters, and Daniel Pauly (Pauly and Pitcher, 2000) with the models of this study may offer a way for future research to evaluate both the present models (with explicit, analytically tractable hypotheses about allometry, links, and fluxes) and the Ecosim and Ecopath models (programmed packages where the core assumptions and their implementation may be less transparent).

C. Weaknesses

If the production (increase in body mass per unit time plus reproduction) of an individual in species i of body mass M is pM^α and the population of species i contains $N(M)$ individuals of body mass M, then the species abundance is $N_i = \int_0^\infty N(M)dM$ and the average body mass is $M_i = (1/N_i)\int_0^\infty M * N(M)dM$. The aggregated production of all individuals of species i is then $\int_0^\infty N(M)pM^\alpha dM \leq pN_iM_i^\alpha$. The inequality follows from Jensen's inequality, because M^α is a concave function of M when $\alpha < 1$, as assumed. The inequality is strict as long as there are individuals of at least two different body sizes in species i. An identical argument and inequality apply to the population consumption of species i. If the allometric functions apply to individuals, then the allometric functions for population production and population consumption must overstate the production and consumption aggregated over all individuals, respectively. This overstatement has been ignored. The significance of intraspecific variation in body mass for interspecific allometric estimates was independently recognized and has been analyzed significantly further by Savage (2004).

Another weakness of the study, as already mentioned, is the lack of empirical flux measurements to compare with the predictions of the flux models.

XII. CONCLUSIONS

All models presented (except the EF model) performed well enough on the tests done to be considered plausible. The MA was the most plausible because it performed slightly better than the other models and because it is conceptually simpler, more easily applied, and more readily visualized. A plot of an (M, N)-web on $\log(M)$ versus $\log(N)$ coordinates, with equi-production and equiconsumption lines based on the standard allometric formulas for production and consumption, yielded more visual information about the relative magnitude of fluxes under the MA model than under any other model. The MA and BR models were also the easiest to use in conjunction with the theory of sampling effort developed in the section on theory. Whether the MA model produces realistic flux estimates can be determined only by empirical measurements of flux in a real ecosystem.

From the perspective of this study, ideal community food-web data should include:

1. Time and location of capture of each individual organism, to test whether the system is temporally and spatially homogeneous.
2. M measurements for individual organisms, and N measurements for each unit of species sampling effort.
3. Age measurements or estimates for individual organisms.
4. Individual organism gut content analysis for nonbasal species, using visual analysis or PCR.
5. Individual organism stable isotope measurements and stoichiometry of at least C, N, P.

This list is not exhaustive and may not be entirely practical under all circumstances.

The choice of ecosystem and the sampling design should be made in light of the following considerations.

1. The ecosystem should be chosen to obtain a food web that is as nearly complete as possible with minimal sampling effort. This choice should be informed by the sampling theory of this study.
2. A system should be chosen for which an initial assumption of spatial homogeneity is reasonable, or separate sampling designs and evaluations

should be constructed for clearly distinguishable major spatial components (e.g., pelagic versus benthic versus littoral, below ground versus above ground).

3. Sampling should be done speedily enough to justify assuming that the ecosystem has not changed much during the sampling, or separate sampling designs and evaluations should be constructed for clearly distinguishable major temporal components (e.g., major seasonal differences or different precipitation regimes).

4. Within each spatial or temporal component, sampling should be intensive enough to reach the point of diminishing returns (i.e., until species yield-effort curves and link yield-effort curves nearly cease to increase with additional sampling effort).

5. Sampling should be continued until M and N data for each species reach a specified coefficient of variation. The value of 10% was used in Tuesday Lake.

Additional desiderata for food-web data were discussed in Cohen *et al.* (1993).

ACKNOWLEDGMENTS

We thank James Brown, Hal Caswell, Donald L. DeAngelis, James Gillooly, Joshua Weitz, and two anonymous referees for constructive suggestions. We acknowledge with thanks the support of U.S. National Science Foundation grant DEB 9981552, the assistance of Kathe Rogerson, and the hospitality of Mr. and Mrs. William T. Golden during this work.

REFERENCES

Adams, S.M., Kimmel, B.L. and Plosky, G.R. (1983) Sources of organic matter for reservoir fish production: A trophic dynamics analysis. *Can. J. Fish. Aquat. Sci.* **40**, 1480–1495.

Bersier, L.-F., Banasek-Richter, C. and Cattin, M.-F. (2002) Quantitative descriptors of food web matrices. *Ecology* **83**, 2394–2407.

Cohen, J.E., Beaver, R., Cousins, S., DeAngelis, D., Goldwasser, L., Heong, K.L., Holt, R., Kohn, A., Lawton, J., Magnuson, J., Martinez, N., O'Malley, R., Page, L., Patten, B., Pimm, S., Polis, G., Rejmànek, M., Schoener, T., Schoenly, K., Sprules, W.G., Teal, J., Ulanowicz, R., Warren, P., Wilbur, H. and Yodzis, P. (1993) Improving food webs. *Ecology* **74**, 252–259.

Cohen, J.E., Jonsson, T. and Carpenter, S. (2003) Ecological community description using the food web, species abundance, and body size. *PNAS* **100**, 1781–1786.

Cohen, J.E. and Luczak, T. (1992) Trophic levels in community food webs. *Evol. Ecol.* **6**, 73–89.

Cousins, S. (2003) Measuring the ability of food to do work in ecosystems. *Food Webs 2003*. Giessen Germany, November 2003.

DeNiro, M.J. and Epstein, S. (1981) Influence of diet on the distribution of nitrogen isotopes in animals. *Geochimica Cosmochimica Acta* **45**, 341–351.

deRuiter, P.C., Neutel, A.-M. and Moore, J.C. (1995) Energetics, patterns of interaction strengths, and stability in real ecosystems. *Science* **269**, 1257–1260.

Emmerson, M.C. and Raffaelli, D. (2004) Predator-prey body size, interaction strength and the stability of a real food web. *J. Anim. Ecol.* **73**, 399–409.

Gaedke, U. and Straile, D. (1994) Seasonal changes of trophic transfer efficiencies in a plankton food web derived from biomass size distributions and network analysis. *Ecol. Model.* **75**, 435–445.

Greenstone, M.H. and Shufran, K.A. (2003) Spider predation: Species specific identification of gut contents by polymerase chain reaction. *J. Arachnol.* **31**, 131–134.

Jarque, C.M. and Bera, A.K. (1987) A test for normality of observations and regression residuals. *Int. Stat. Rev.* **55**, 163–172.

Jennings, S., Pinnegar, J.K., Polunin, N.V.C. and Warr, K.J. (2002a) Linking size-based and trophic analyses of benthic community structure. *Mar. Ecol. Prog. Ser.* **226**, 77–85.

Jennings, S., Warr, K. and Mackinson, S. (2002b) Use of size-based production and stable isotope analyses to predict trophic transfer efficiencies and predator-prey body mass ratios in food webs. *Mar. Ecol. Prog. Ser.* **240**, 11–20.

Jonsson, T., Cohen, J.E. and Carpenter, S.R. (2005) Food webs, body size and species abundance in ecological community description. *Adv. Ecol. Res.* **36**, 1–84.

Kling, G.W., Fry, B. and O'Brien, W.J. (1992) Stable isotopes and planktonic trophic structure in arctic lakes. *Ecology* **73**, 561–566.

Kokkoris, G.D., Troumbis, A.Y. and Lawton, J.H. (1999) Patterns of species interaction strength in assembled theoretical competition communities. *Ecol. Lett.* **2**, 70–74.

Kooijman, S.A.L.M. (2000) *Dynamic Energy and Mass Budgets in Biological Systems*. Cambridge University Press, Cambridge.

Lilliefors, H.W. (1967) On the Kolmogorov-Smirnov test for normality with mean and variance unknown. *J. Am. Stat. Assoc.* **62**, 399–402.

Martinez, N.D. (1991) Artifacts or attributes: Effects of resolution on the Little Rock Lake food web. *Ecol. Monogr.* **61**, 367–392.

Martinez, N.D., Hawkins, B.A., Dawah, H.A. and Feifarek, B.P. (1999) Effects of sampling effort on characterization of food web structure. *Ecology* **80**, 1044–1055.

McCann, K., Hastings, A. and Huxel, G.R. (1998) Weak trophic interactions and the balance of nature. *Nature* **395**, 794–798.

Minagawa, M. and Wada, E. (1984) Stepwise enrichment of 15N along food chains: further evidence, and the relation between 15N and animal age. *Geochimica Cosmochimica Acta* **48**, 1135–1140.

Moore, J.C., deRuiter, P.C. and Hunt, H.W. (1993) Influence of productivity on the stability of real and model ecosystems. *Science* **261**, 906–908.

Morales, M.E., Wesson, D.W., Sutherland, I.W., Impoinvil, D.E., Mbogo, C.M., Githure, J.I. and Beier, J.C. (2003) Determination of *Anopheles gambiae* larval DNA in the gut of insectivorous dragonfly (Libellulidae) nymphs by polymerase chain reaction. *J. Am. Mosquito Cont. Assoc.* **19**, 163–165.

Neutel, A.-M., Heesterbeek, J.A.P. and de Ruiter, P.C. (2002) Stability in real food webs: weak links and long loops. *Science* **296**, 1120–1123.

Paine, R.T. (1992) Food-web analysis through field measurement of per capita interaction strength. *Nature* **355**, 73–75.

Pauly, D., Christensen, V., Guenette, S., Pitcher, T.J., Sumaila, R.S., Walters, C.J., Watson, R. and Zeller, D. (2002) Toward sustainability in world fisheries. *Nature* **416**, 689–695.

Pauly, D. and Pitcher, T.J. (Eds). (2000) Methods for assessing the impact of fisheries on marine ecosystems of the North Atlantic. Fisheries Centre Research Reports 8(2). Available at:http://www.seaaroundus.org/report/methodF.htm.

Peters, R.H. (1983) *The Ecological Implications of Body Size*. Cambridge University Press, New York.

Peterson, B.J. and Fry, B. (1987) Stable isotopes in ecosystem studies. *Ann. Rev. Ecol. Syst.* **18**, 293–320.

Phillipson, J. (1966) *Ecological Energetics*. In: *Institute of Biology Studies in Biology no. 1*. Edward Arnold Ltd, London.

Post, D.M. (2002) Using stable isotopes to estimate trophic position: models, methods, and assumptions. *Ecology* **83**, 703–718.

Post, D.M., Pace, M.L. and Hairston, N.G. (2000) Ecosystem size determines food chain length in lakes. *Nature* **405**, 1047–1049.

Raffaelli, D.G. and Hall, S.J. (1996) Assessing the relative importance of trophic links in food webs. In: *Food webs: Integration of patterns & dynamics* (Ed. by G. A. Polis and K.O. Winemiller), pp. 185–191. Chapman & Hall, New York.

Reuman, D.C. and Cohen, J.E. (2004) Trophic links' length and slope in the Tuesday Lake food web with species body mass and numerical abundance. *J. Anim. Ecol.* **73**, 852–866.

Rott, A.S. and Godfray, H.C.J. (2000) The structure of a leafminer-parasitoid community. *J. Anim. Ecol.* **69**, 274–289.

Savage, V.M. (2004) Improved approximations to scaling relationships for species, populations, and ecosystems across latitudinal and elevational gradients. *J. Theor. Biol.* **227**, 525–534.

Schoener, T.W. (1989) Food webs from the small to the large: Probes and hypotheses. *Ecology* **70**, 1559–1589.

Ulanowicz, R.E. (1984) Community measures of marine food networks and their possible applications. In: *Flows of Energy and Materials in Marine Ecosystems* (Ed. by M.J.R. Fasham), pp. 23–47. Plenum, London.

Winemiller, K.O. (1990) Spatial and temporal variation in tropical fish trophic networks. *Ecol. Monogr.* **60**, 331–367.

Woodward, G., Spears, D.C. and Hildrew, A.G. (2005) Quantification and resolution of a complex, size-structured food web. *Adv. Ecol. Res.* **36**, 85–136.

Yodzis, P. (1989) *Introduction to Theoretical Ecology*. Harper & Row, New York, USA.

Zanden, M.J.V. and Rasmussen, J.B. (1999) Primary consumer δ-^{13}C and δ-^{15}N and the trophic position of aquatic consumers. *Ecology* **80**, 1395–1404.

Index

Advances in Ecological Research
Volume 1–36

Cumulative List of Titles

Aerial heavy metal pollution and terrestrial ecosystems, **11**, 218

Age determination and growth of Baikal seals (*Phoca sibirica*), **31**, 449

Age-related decline in forest productivity: pattern and process, **27**, 213

Analysis and interpretation of long-term studies investigating responses to climate change, **35**, 111

Analysis of processes involved in the natural control of insects, **2**, 1

Ancient Lake Pennon and its endemic molluscan faun (Central Europe; Mio-Pliocene), **31**, 463

Ant-plant-homopteran interactions, **16**, 53

Arrival and departure dates, **35**, 1

The benthic invertebrates of Lake Khubsugul, Mongolia, **31**, 97

Biogeography and species diversity of diatoms in the northern basin of Lake Tanganyika, **31**, 115

Biological strategies of nutrient cycling in soil systems, **13**, 1

Bray-Curtis ordination: an effective strategy for analysis of multivariate ecological data, **14**, 1

Breeding dates and reproductive performance, **35**, 69

Can a general hypothesis explain population cycles of forest lepidoptera?, **18**, 179

Carbon allocation in trees; a review of concepts for modeling, **25**, 60

Catchment properties and the transport of major elements to estuaries, **29**, 1

Coevolution of mycorrhizal symbionts and their hosts to metal-contaminated environment, **30**, 69

Conservation of the endemic cichlid fishes of Lake Tanganyika; implications from population-level studies based on mitochondrial DNA, **31**, 539

The cost of living: field metabolic rates of small mammals, **30**, 177

A century of evolution in *Spartina anglica*, **21**, 1

The challenge of future research on climate change and avian biology, **35**, 237

Climate influences on avian population dynamics, **35**, 185

The climatic response to greenhouse gases, **22**, 1

Communities of parasitoids associated with leafhoppers and planthoppers in Europe, **17**, 282

Community structure and interaction webs in shallow marine hardbottom communities: tests of an environmental stress model, **19**, 189

The decomposition of emergent macrophytes in fresh water, **14**, 115

Delays, demography and cycles; a forensic study, **28**, 127

Dendroecology; a tool for evaluating variations in past and present forest environments, **19**, 111